NOTEBOOK ON STRUCTURAL BIOLOGY TECHNIQUES

Study guide for Practitioners

NOTEBOOK ON STRUCTURAL BIOLOGY TECHNIQUES

Study guide for Practitioners

Ipsita Banerjee
Lund University, Sweden

Editors

Rabin Banerjee
Formerly, S.N. Bose National Centre for Basic Sciences, Kolkata, India

Jayati Sengupta
Indian Institute of Chemical Biology, Kolkata, India

World Scientific

NEW JERSEY • LONDON • SINGAPORE • BEIJING • SHANGHAI • TAIPEI • CHENNAI

Published by

World Scientific Publishing Co. Pte. Ltd.
5 Toh Tuck Link, Singapore 596224
USA office: 27 Warren Street, Suite 401-402, Hackensack, NJ 07601
UK office: 57 Shelton Street, Covent Garden, London WC2H 9HE

Library of Congress Control Number: 2024943385

British Library Cataloguing-in-Publication Data
A catalogue record for this book is available from the British Library.

NOTEBOOK ON STRUCTURAL BIOLOGY TECHNIQUES
Study guide for Practitioners

ISBN 978-981-12-9652-9 (hardcover)
ISBN 978-981-12-9653-6 (ebook for institutions)
ISBN 978-981-12-9654-3 (ebook for individuals)

For any available supplementary material, please visit
https://www.worldscientific.com/worldscibooks/10.1142/13941#t=suppl

Desk Editor: Carmen Teo Bin Jie

Typeset by Stallion Press
Email: enquiries@stallionpress.com

Preface

The recent plethora of interdisciplinary subjects seems to confirm the brilliant and charismatic physicist Richard Feynman's views that science knows no boundaries. Its division into physics, chemistry, biology, mathematics etc. was done by humans for our own convenience and purpose. Nature is least bothered by these divisions. Time and again it has been proven that a holistic approach to science is much more effective than compartmentalising it.

In many ways the subject of this book is a nice example of the above philosophy. Structural biology has been enriched by concepts and ideas from biology, physics, chemistry, mathematics and computer science. Interestingly, recently artificial intelligence (AI) and/or machine learning (ML) have also intruded in this field.

Among the various approaches cryo-EM (cryogenic electron microscopy) has proved very effective in producing high (near atomic scale) resolution micrographs. Despite its increasing inportance and relevance, there is no simple easy-to-read material on this subject, especially for students. This caveat is filled, albeit to a small extent, by the present primer, which concentrates on the methods aspect.

The primer has certain unique features. It is the work of a Ph.D. candidate Ipsita Banerjee, at Lund University who, just at the point of completion, passed away suddenly on 7th December 2022. Coming from far-off India with no previous experience in experimental science, she landed in the harsh climate of Sweden amongst unfamiliar people. Her initial project, suggested by her mentor Prof. Derek Logan, was based on crystallographic studies, but did not yield fruitful results. At this point her mentor shifted the focus to cryo-EM.

Initially, Ipsita was at a loss. She was completely alone with no one to help her at the University, since nobody knew the subject. Naturally, there was little or no progress. However she was the last person to give up or admit defeat. She started learning cryo-EM completely on her own and mastered it to a level that left her colleagues and mentors dumbfounded. She was recognised as a pioneer and her seminal contribution was instrumental in establishing cryo-EM at Lund University.

She wrote an extensive 'Introduction' to her thesis, covering the methods aspect of structural biology, especially cryo-EM. It is considerably longer than what a Swedish Ph.D. candidate would normally write. After she passed away, her thesis notes were suitably compiled, edited and placed in a logical manner. Finally, the constant support and guidance of the Editor of WSPC (World Scientific Publishing Company) Carmen Teo Bin Jie, eventually resulted in the present primer. We hope that young researchers, especially graduate students, would find it useful and a practical source of information.

R. Banerjee and J. Sengupta (Editors)

Contents

Chapter 1

X-ray Crystallography

X-ray crystallography helps us to determine the atomic and molecular structure of a crystal experimentally. The crystalline structure present in a crystal causes X-ray beams entering the crystal to diffract in many directions. Protein X-ray crystallography is a method which is used to obtain the three-dimensional structure of a particular protein by X-ray diffraction of its crystallised form. This three-dimensional structure is essential for determining the protein's function (Figure 1).

A brief history of X-ray crystallography

William Hallowes Miller proved in 1839, in no uncertain terms, that a crystal is an ordered lattice. In other words it exhibits periodicity that manifests the specific ordering in the crystal. However, it was much later that clinching evidence was given which established

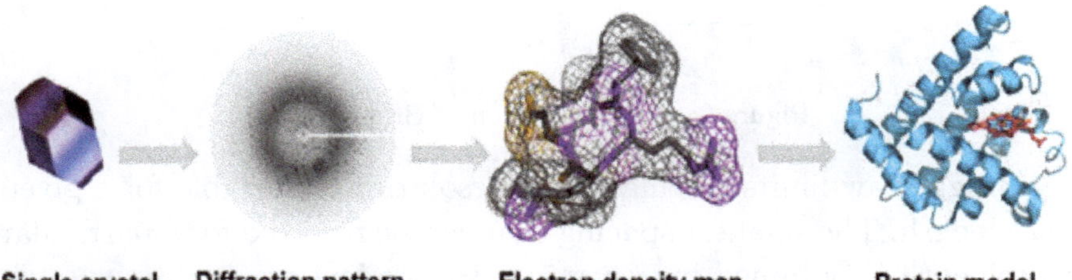

Figure 1: Creative biostructure (Comparison of crystallography, NMR and EM).[1]

[1]Figure taken from: https://www.x-mol.com/groups/Dong/news/1402in.

crystallography as a science. It began with the discovery of X-rays in 1895. Subsequently, it was shown by Max von Laue in 1912 that these X-rays undergo diffraction. The year 1912 may, therefore, be taken as the beginning of crystallography. Soon after, the father and son duo William Henry Bragg and William Lawrence Bragg shared the 1915 Nobel Prize in physics for discovery of Bragg's law:

$$n\lambda = 2d \sin\theta$$

which relates an X-ray diffraction pattern to the three-dimensional structure of a crystal. Here $\boldsymbol{\lambda}$ is the wavelength of light that is scattered at an angle $\boldsymbol{\theta}$ from a lattice whose atoms are spaced at a distance $\boldsymbol{d}$ between the parallel planes and $\boldsymbol{n}$ is a number that indicates the order of diffraction.

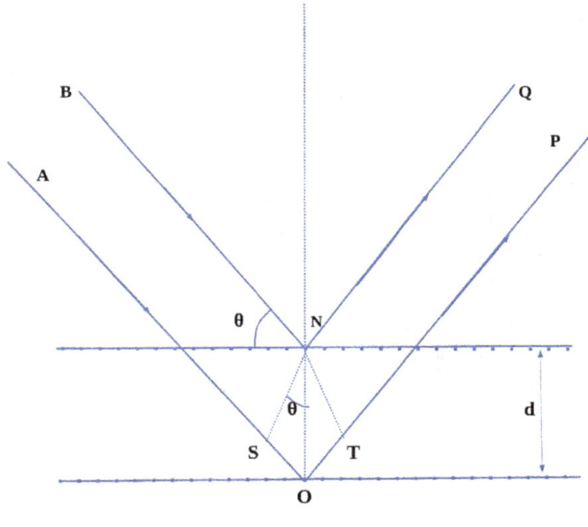

Figure 2: Demonstration of Bragg's law.

Bragg's law limits the maximum resolution achievable for a given wavelength. The smallest spacing that can be resolved for a particular wavelength is obtained by rewriting Bragg's law as,

$$d = \frac{n\lambda}{2 \sin\theta}$$

The smallest value of the numerator occurs for $n = 1$ and the largest in the denominator occurs for $\sin\theta = 1\,(\theta = 90^\circ)$ when waves simply go straight and turn back. Thus the smallest achievable value of d is $\frac{\lambda}{2}$.

It is simple to prove Bragg's law from the above diagram (Figure 2). There is a path difference between the two rays BNQ and AOP that is given by OS+OT which simplifies to $2d \sin\theta$. When this equals an integral multiple of the wavelength we have constructive interference and we reproduce Bragg's law given above.

It is useful to point out that while both refraction and diffraction bend light rays, there is an important difference. In the case of refraction, waves of high frequency or low wavelength are deviated the most while for diffraction, it is just the opposite. The genesis of this is contained in the distinct reasons for the happening of these two phenomena. Refraction obeys Snell's law while diffraction follows Bragg's law.

The subject of X-ray crystallography has a chequered history leading to innumerable Nobel prizes over the years. As specific examples, Dorothy Crowfoot Hodgkin who solved the structures of small molecules chloesterol, penicillin and Vitamin B12 was awarded this prize for Chemistry in 1964. A couple of years earlier, in 1962, the Nobel prize for Chemistry was given to Max Perutz and John Kendrew for their work on sperm whale myoglobin. David Chilton Phillips in 1965 cracked the first structure of an enzyme, iysozyme. The early 70s saw the birth and development of the Research Collaboratory for Structural Bioinformatics Protein Data Bank (PDB). The PDB started with 13 structures in 1976 and has since grown to the "single worldwide archive of structural data of biological macromolecules".[2]

Protein crystallization

The crystallization of proteins, nucleic acids and large biological complexes like viruses depends on the preparation of a solution that is super-saturated in the macromolecule but also exhibits conditions that do not significantly disturb its natural state. Supersaturation

[2]Physics Libre Texts, X-ray crystallography, Univ of California Davis.

is usually achieved through the addition of mild precipitating agents like neutral salts or polymers and manipulation of various parameters that include temperature, ionic strengths and pH. The different approaches that have been developed to effect and promote crystallization are as follows: vapour diffusion, dialysis, batch and liquid-liquid diffusion.[3]

A two-dimensional phase diagram characterizing the different stable states (liquid, crystalline, precipitate) as a function of two crystallization variables shows the build-up of a crystal. When the concentration of a protein solution exceeds its solubility limit, a supersaturated solution is obtained. There are three regions in this phase: one is an extremely high supersaturation "precipitation", another is an intermediate supersaturation ("labile") where growth and nucleation simultaneously occur and finally, the area of lower supersaturation where only growth happens (Figure 3). Their boundaries are obscured because these zones are connected to kinetic quantities. Possibly the most suitable and convenient approach

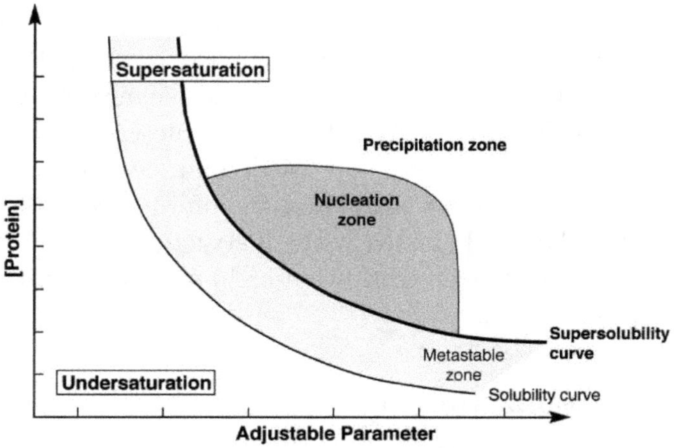

Figure 3: Phase structure of protein solution.[4]

[3]Introduction to protein crystallization; Alexander McPherson, Jose A. Gavira; *Acta Crystallogr F Struct Biol Commun.* 2014 Jan 1; 70(Pt 1): 2–20.

[4]Figure taken from: https://www.semanticscholar.org/paper/Methods-for-separating-nucleation-and-growth-in-Chayen/06ea1d3acd9b6bda66fe4f44de37f223da87b7b3.

is to force nucleation at the minimum level of supersaturation, just inside the labile zone. After the formation of the nuclei, the concentration of protein in the solution slowly diminishes. This triggers the system into the metastable zone where growth takes place slowly.

In spite of the fact that protein crystallization is a key step of X-ray crystallography, the details and finer points of this process are still not completely known. Presently, there are no definite approaches available to ensure that ordered three-dimensional crystals will be found.[5]

Crystal optimization

The first step in protein crystallography is obtaining a protein crystal that diffracts in the X-ray beam. A new protein poses problems in determining its conditions for crystallization. Initially, screening trials are established. The next step is to expose the protein to different agents so that appropriate 'hits' or 'leads' are obtained. These are helpful in determining conditions favourable for crystallization. For example, leads are provided by crystals, crystalline precipitate and phase separation. After the lead is found, there are a couple of methods in which optimization can be performed. The most popular approach is to appropriately alter the protein concentration, type and precipitate concentration, pH or temperature that enables the fine tuning of crystallization conditions. Alternatively, one proceeds by adding additives for a second screening with a more defined range of conditions.

Classic screening methods

There are many screening methods. However, the sparse-matrix screens (for example the Crystal Screen from Hampton Research) appear to be the most favoured ones. Initially, the screens had to depend on a set of conditions that had earlier proved their success in crystallization. They are still very popular and commonly used.

---

[5] An overview of biological macromolecule crystallization. Russo Krauss, *Int J Mol Sci.* 31 May 2013; 14(6): 11643–11691.

However there has been a paradigm shift. Current trends in sparse matrix screens depend on recently found properties of precipitants and additives. The crystallization parameter space is sampled in a balanced, logical manner by the use of systematic screens. Such screens can be suitably fabricated and there are also commercially available products (like Clear Strategy Screens from Molecular Dimensions and the JBScreen PEG/Salt from Jena Bioscience).

Choosing the method of crystallization

There are different methods of crystallization that can be used:

Microbatch experiments

Crystallization trials are not needed in microscale batch experiments. Low density paraffin oil (0.87 g/ml) is employed for incubation. Since the aqueous crystallization drops are normally denser than the paraffin oil, they remain submerged below the oil. This acts as a protective shield from evaporation and airborne impurities.

Vapor diffusion

Vapor diffusion is the most widely used technique of protein crystallization. A hanging or sitting drop constitutes the protein solution. This equilibrates against a container with crystallization agents that are kept either at higher or lower concentrations than in the drop. 24-well plates are generally employed to perform the trials. All wells are filled with 0.5–1 ml reservoir solution and then sealed with either a plastic coverslip or siliconized glass. The sitting drop has a support that is fitted to the wells, while the hanging drops dangle from the cover slip (Figures 4 and 5).

Dialysis

The protein is initially secluded from the precipitant solution in this approach. This is achieved using a semi-permeable membrane. This ensures that the precipitant solution mixes slowly with the protein molecules.

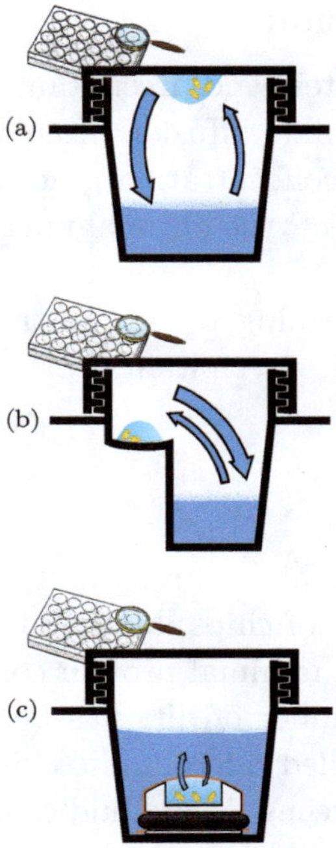

Figure 4: Sitting drop method. https://commons.wikimedia.org/wiki/File:CrystalDrops.svg.

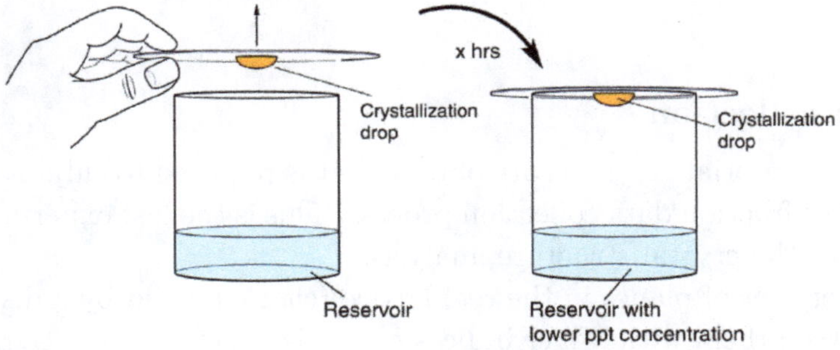

Figure 5: Hanging drop method.[6]

[6]Figure taken from:https://www.semanticscholar.org/paper/Methods-for-separating-nucleation-and-growth-in-Chayen/06ea1d3acd9b6bda66fe4f44de37f223da87b7b3. Fig: Hanging drop method: http://soft-matter.seas.harvard.edu/index.php/Vapor_Diffusion_Method.

Free interface diffusion

In this approach the protein and precipitant solutions are juxtaposed such that there is a mild diffusion from one to the other. This induces a time varying concentration gradient. Now the system has the potential to 'self-select' the optimal nucleation and growth super saturation level.

As a fine-tuning screening technique this is extremely powerful when suitable additives, precipitants, pH and buffers are already established.

Microfluidics

Apart from other uses, high throughput crystallization screening rely a lot on this variety of chips. Microfluidics have the remarkable property to ensure both minimal protein consumption and maximal speed. An enhanced number of hits has been found in experiments using microvalve controlled microfluidic chips as compared to standard or conventional screens. Microfluidics are very useful but their flip sides are expensive hardwares and consumables.

Methods of choice for beginners are microbatch, where the crystallization parameters can be kept under best control, and vapour diffusion, for its wide-range sampling of the crystallization parameter space.

Data collection

Once appropriate crystals are obtained, it is required to submit them to the diffraction data collection process. This is the last experimental stage of the crystal structure analysis.

Every set of planes in the real lattice (characterized by a distance d between them with Miller indices h, k, l) is normal to the reciprocal lattice vector defined as,

$$D^* = h.a^* + k.b^* + l.c^*$$

where the distance between planes in the reciprocal lattice is given by d^*, where $d^* = 1/d$. For more details see, Mathematical Supplement on Miller indices.

Every set of planes correspond to a X-ray reflection. If we measure all reflections from all planes F(hkl), the electron density distribution can be calculated through a summation:

$$\rho(r) = 1/V \sum F(\text{hkl}) \exp(-2\pi i(hx + ky + lz))$$

where V is the volume of the unit cell. This is a Fourier synthesis.

Data collection usually takes place at synchrotrons which have the advantages of having high flux, highly parallel beam, small focal spot (down to 5–50 micron) and variable wavelength (about 0.7–2.0 Å). Crystals are rotated by 0.1 deg. Many pictures are collected to obtain the full "reciprocal space".

We need to obtain the electron density distribution,

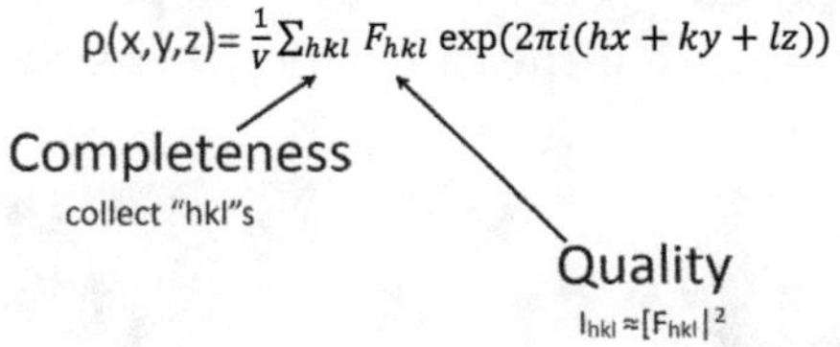

Improving the quality can be achieved by increasing I/sigma. Increasing intensity (I) can be obtained by either increasing the brilliance of the beam or growing bigger crystals. Reducing sigma, leading to improved quality, is achieved by reducing background noise or measuring on longer time scales. Completeness of data is necessary as there would otherwise be a possibility of missing high resolution data, lack of detail and ripples. During data collection as many reflections as possible from various planes are required. How much of the theoretically possible reflections we actually collect is given by the completeness. Collecting more data usually gives an option to discard, tailor or merge data in post collection processing. The more symmetry related observations of each unique reflection that we merge together (multiplicity) the more statistically reliable are the intensities of the unique reflections. However, getting good multiplicity comes at a cost of radiation damage (i.e. the crystal dies with time) (Data collection study material, 2016, Marjolein.)

Data processing

Indexing

One of the first steps in data processing is indexing. There are two separate pieces of information that can be obtained in the reflections of the diffraction images. The first is the geometrical arrangement of the reflections, which gives all the information about the crystal lattice and symmetry of the crystal. The second piece of information comes from the intensity of reflection that gives partial information about the lattice content. Unfortunately, the information that we record is partial — we lack the phases. During the step of indexing, after locating the spots, they have to be properly identified with their Miller indices which are some integers (h, k, l) (Figure 6). From this information, the crystal geometry should be properly obtained. This is necessary for integrating the intensities correctly.

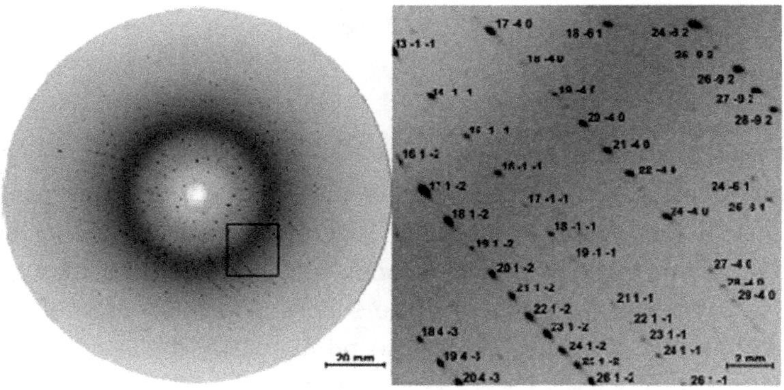

Figure 6: Labelled diffraction pattern with miller indices (my.yetnet.ch).

Integration

The actual experimental data of a crystallography experiment are the intensities of the reflections. Thus, in data processing, integration is a basic step. An intensity value (I) and σ, that accounts for the reliability of the intensity estimate in the form of standard deviation of a normal distribution around the observed value, are calculated and stored. However, many factors and errors may creep in that have to be kept in mind. For example, the backstop of the detector can conceal the reflections. Also, the signal can get saturated and a large percentage of it can be hidden in the background. Wrong values and errors such as these have to be eliminated from the analysis. This is normally done by the help of standard integration programs like MOSFLM and HKL2000.

Scaling

During scaling, the integrated values of the different reflections which were collected in the diffraction experiment are combined into one set of structure factors and normalized which is also done according to symmetry. The structure factor represents the resultant amplitude and phase of scattering of all the electron density distribution of one unit cell. The amplitude is calculated as the number of times it is greater than the amplitude of scattering from a single electron. The intensity is proportional to the modulus square of the structure factor which is effectively equal to the square of the resultant amplitude.

Phasing

Phase problem in X-ray crystallography

X-rays scattered from each point in a crystal along a specific direction interfere among themselves. This generates a diffraction pattern that depends on the arrangement of atoms in the crystal. Hence, analysis of the diffraction pattern can therefore allow the arrangement of atoms to be deduced. The intensity of radiation scattered in any specific direction from the crystal depends on whether X-rays scattered along that direction interfere constructively or destructively. This also depends on the position and spacing of electron density features (in particular atoms) within the crystal.

The crystal is an ordered lattice of identical molecules. Hence every feature of the electron density will be periodically recovered at regular intervals, i.e. there is a definite periodicity in the electron density map. The unit cell is the basic recurring item from which a crystal is formed. Among all directions that scattering occurs, a special case is the direction along which the unit cell is repeated.

The repetitions of the unit cells will reinforce and strengthen the scattering along this particular direction. Scattering along all the other directions will be feeble. Consequently, the full diffraction pattern of the crystal contains spots that form a 3-dimensional lattice with reciprocal directions and spacings from the real lattice (for more details see Mathematical Supplement on Miller indices). An individual diffraction image which is found from a single crystal orientation is a 2D slice through this pattern. This is known as reciprocal lattice. The lattice spacing here is the reciprocal of the lattice spacing in a real (direct) lattice. In other words if the atoms are more densely packed in a real lattice, they appear more loosely packed in the reciprocal lattice.

The diffraction spots are of varying intensities. They depend on the changing patterns of the electron density in the individual unit cells. The intensities of the diffraction spots and the density within the unit cell are connected by a mathematical formula known as Fourier transform that was mentioned earlier (For more details on Fourier transforms, see Mathematical Supplement). Thus the knowledge of the contents of the unit cell enables one to construct the diffraction pattern. The reverse is also true, knowledge of the scattering from a crystal determines the contents of the unit cell. This knowledge, rather than that of the full crystal, is sufficient. Indeed Fourier transformations ensure that only the diffraction spots, and not their intervening space, are important. Each of these spots is characterized by a single wave indicating the magnitude and relative phase of the X-rays scattered along that direction. This wave is mathematically represented as a structure factor with a specific amplitude and phase. The problem arises in the fact that an experiment only yields the amplitude of the diffraction pattern but not the phase. This is not surprising since the amplitude is connected with the real part and the phase with the imaginary part of the structure factor. This lack of phase information is referred as the phase problem in X-ray crystallography.[7]

---

[7]Phase Problem in X-ray Crystallography and its Solution, Kevin Cowtan, York, UK; Encyclopedia of Life Sciences, 2001; Macmillan Publishers Ltd, Nature Publishing Group/www.els.net.

The problem in dealing with phase is a major issue in X-ray crystallography and it's one of the hurdles which make structure determination quite difficult. It stems from the fact that only the intensities but not the phases of the electromagnetic waves are captured by the detectors. A specific wave with a definite amplitude and phase is linked to a reflection spot on the diffraction pattern or a structure factor. Since amplitude is the square root of the intensity, it can be easily obtained from the available data, however the phase is lost. The problem arises because crucial information related to the abstraction of the electron density distribution in the crystal is buried in the phases. Thus there is a concerted attempt to solve this phase problem and a substantial amount of crystallography is devoted to it. A structure factor can be represented as a two dimensional vector or a complex number F_{hkl} with an amplitude $|F_{hkl}|$ and a phase ϕ_{hkl}. This representation is called an Argand diagram, shown below (Figure 7).

The structure factor F_{hkl} is expressed in terms of its real and imaginary components as:

$$F_{hkl} = |F_{hkl}| \exp(i\phi) = |F_{hkl}|(\cos\phi + i\sin\phi)$$

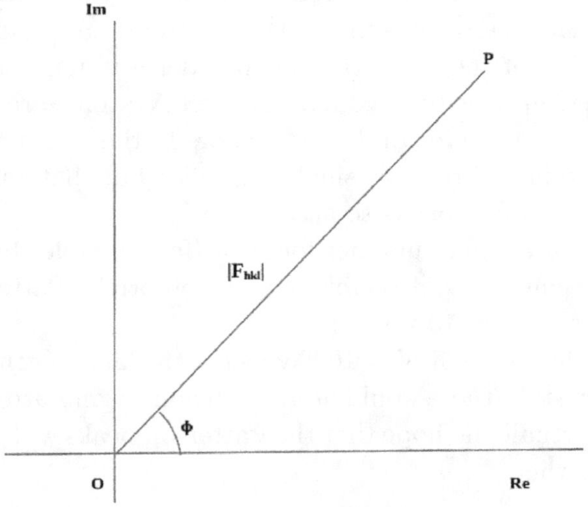

Figure 7: Argand diagram.

Retrieving the lost phases (solving the phase problem)

Patterson methods

A notion of the electron density can be obtained by performing an inverse Fourier transform of the structure factors (including both amplitudes and phases). Since the measured experimental data gives only the intensities (i.e. squared amplitudes) one might be tempted to ask what would happen if the Fourier transform was taken for intensities only. This was precisely the question asked by Patterson. The resultant map, that goes by the name of Patterson function or Patterson map, has plenty of attractive points.

The Patterson function gives a picture of the interatomic vectors. Consider an electron density for atom 1 which has a maximum at position x_1 and likewise for atom 2 at position x_2. Then the Patterson map will have maxima at positions given by $x_2 - x_1$ and $x_1 - x_2$. Furthermore, this Patterson map will have a peak height that is proportional to the product of the heights of the two peaks in the electron density map. From the observed Patterson peaks one may obtain the original configuration of the atoms. However, this is practically feasible if the number of atoms is small. Known as deconvolution of the Patterson map, the method fails for larger molecules. The reason is simple. For N atoms in a unit cell, and high resolution of the data, the electron density map will reveal N distinct electron density maxima. Since all N atoms are involved in the configuration, there will be N^2 vectors in the relevant Patterson map. The origin will reveal a single big spike, but that will still leave $N(N-1)$ non origin peaks to account for.

Take a typical small number for $N =$ (for example, 10).

Then we have a controllable set of non-origin Patterson peaks because $N(N-1) = 10 \times 9 = 90$.

On the other hand, if $N = 1000$ which is the kind of value expected in protein crystals, there would be 999,000 non origin Patterson peaks. So there is virtually no hope that the Patterson peaks will get resolved from one another.[8]

[8]structmed.cimr.cam.ac.uk.

Molecular replacement

Molecular replacement is a technique for solving the phase problem from the knowledge of earlier solved protein structure that is similar to the unknown structure from which the diffraction data is obtained. It is a familiar approach for recovering the lost phases. This approach is employed when there is a valid model for an appreciably big fraction of the structure in the crystal. The method is quite straightforward if the model is, to a good extent, complete and shares a minimum 40% sequence identity with the unknown structure. Conversely, it is increasingly difficult to work with this technique as the model becomes less complete or shares a diminished sequence identity.

The model structure has to be kept in the proper orientation and position in the unknown unit cell in order to apply molecular replacement. The orientation of the molecule is fixed by choosing six quantities in the unit cell-three rotation angles and three translation parameters. Thus, a single molecule in the asymmetric unit of the crystal leads to a six dimensional problem in the molecular replacement program. Among many options it is run through the program Mr. Bump in the ccp4 software package.

Isomorphous replacement

To solve for the phase of a protein crystal, the complexity of the problem (the number of variables) should be reduced. One way to achieve this is to soak the crystal in a heavy atom solution and compare the diffraction pattern of this crystal to the diffraction pattern of the native crystal. The heavy atoms present in the structure contribute significantly to the diffraction which makes their contribution easy to detect and the difference of the two diffraction patterns can be taken. With direct methods, which work for small molecules, the position of heavy atoms can be calculated.

Anomalous dispersion

Here, in order to obtain information regarding the position within a single crystal, anomalous scattering (or dispersion) of heavy atoms can be utilised. If we deal with wavelength of X-rays corresponding to a transition of the different electron shells in the atoms, then the phase will be altered. The diffraction pattern is influenced by anomalous scattering and hence is distinguishable from the rest. The atoms used here do not need to be as big as the ones for isomorphous replacement. Generally, proteins are used where methionine is replaced by selenomethionine. Normally, data is gathered at peak lambda 1 for single anomalous dispersion (SAD) and also at inflection (lambda 2) and remote lambda 3, 4 for multiple anomalous dispersion (MAD).

Model building, refinement and validation

Model building

In X-ray crystallography the experimental result is the electron density map which has several shortcomings such as resolution limit, measurement errors in structure factor amplitudes as well as uncertainty in phase angles.

Everything that comes after the electron density maps is interpretation. It is here that our best bet is the 3D model leading to a proper interpretation of the electron density with regard to atoms.

Starting model building

If we have no idea about the structure beforehand, we should look out for certain signposts — or identify if our structure contains a known module (e.g. Rossman Fold, TIM barrel) so that we can place all of it at once and then modify it. Usually, we trace the Calpha positions, followed by constructing the main chain, then the side chain. The well-ordered core of the protein is often built first. Then come the N and C terminus, if often disordered, while flexible loops often come in later. It is important to know where we are in the sequence which can be achieved by locating the amino acids Trp, Phe, Tyr, Cys and Met. It is a good idea to build whole secondary structure elements if possible. Helices are usually fitted as rigid bodies, but beta sheets often show more variability and have to be built by hand. The side chains are usually built after the main chain. Model building for both crystallography and cryo-EM maps is done in coot.[9]

Refinement

Real space refinement

Real space refinement (Figure 8) has been successfully applied to solve various protein structures. Its utility lies in the fact that it can incorporate both the magnitude and phase of the observed diffraction pattern in a very simple manner.

It is useful for refining the initial fit by hand to the density. It is employed to minimize the difference between observed and calculated density at points on a predefined grid. The real space R factor is defined as,

$$R_{\text{real-space}} = \sum \left| \frac{\rho_{\text{obs}} - \rho_{\text{calc}}}{\rho_{\text{obs}} + \rho_{\text{calc}}} \right|$$

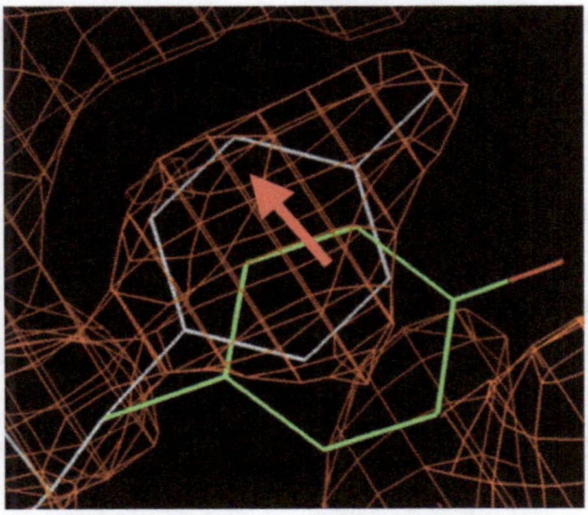

Figure 8: Real space refinement in coot.

[9][Ana Casañal, Bernhard Lohkamp, Paul Emsley] Current developments in Coot for macromolecular model building of Electron Cryo-microscopy and Crystallographic Data. Protein Sci. 2020 Apr; 29(4):1069–1078. doi: 10.1002/pro.3791. Epub 2020 Mar 2.

Real-space R factor: minimise

The atoms are moved in small shifts so that the R factor is minimized.

Real space R can also be calculated per residue for a refined structure to show how well each residue fits the electron density during validation. It is carried out in the program coot for crystallographic and cryo-EM maps. For cryo-EM maps, we also run real space refinement in phenix.

Reciprocal space refinement

It is quite impossible to build an accurate model into the electron density completely by hand. Reciprocal space refinement is done to reduce the manual work and improve the model by minimizing the difference between $|F_o|$ and $|F_c|$ as judged by the model R factor or R model (sometimes called R_{work} or just R)

$$R_{\text{model}} = \frac{\sum_{hkl} ||F_o| - |F_c||}{\sum_{hkl} |F_o|}$$

The phases are not used for the calculation of R model as we don't have observed values for them.

We should use R free to monitor the refinement. The free R value is a statistical quantity for assessing the accuracy of crystal structures.[10] For correct refinement, R free follows R model. For incorrect refinement a large part of the structure was built wrongly. The refinement algorithms can still fit the model parameters to the data to get a low value of R, but R free does not follow. As a result there is a big difference between R model and R free. Reciprocal space refinement is carried out for crystallography maps.

Validation criteria

During model building and also before depositing our final structure we must make sure that our model makes sense. The different criteria we should use are —

[10] Axel T. Brunger, *Nature*, Vol. 355, 30 Jan, 1992.

Geometric and Crystallographic criteria

Ramachandran plot must be excellent and there should be very less regions in the disallowed or forbidden zones. Side chain rotamers and torsion angles must be excellent, and there should be minimal deviations from ideal geometry.

The crystallographic criteria are as follows — R free has been widely used to assess model quality, but recently it has been proven to be poor criteria to judge the quality of a model on its own. The difference R free — R model should be excellent. The real space R and correlation coefficient should be good. This coefficient shows how well our model fits the electron density on a residue-by-residue basis.[11]

[11] [Ana Casañal, Bernhard Lohkamp, Paul Emsley] Current developments in Coot for macromolecular model building of Electron Cryo-microscopy and Crystallographic Data. Protein Sci. 2020 Apr; 29(4):1069–1078. doi: 10.1002/pro.3791. Epub 2020 Mar 2.

Chapter 2

Basics of Electron and Cryo-Electron Microscopy

Before discussing cryo-EM it is useful to mention the salient features of an optical microscope to illustrate their differences. The impact and importance of cryo-EM becomes evident.

An electron microscope (EM) is similar to an optical microscope, except that electrons are used instead of light. The principal advantage is that an EM has considerably greater resolving power than an optical microscope.

Any two points can be properly distinguished provided the wavelength is lesser than the distance between those points. In other words a smaller wavelength has better resolving power than a greater wavelength.

To get a reasonable estimate of wavelengths, consider light of wavelength 5000A^0 $(= 5 \times 10^{-5}\,\text{cm})$ which lies approximately in the middle of the "optical window". An atom has a diameter of 1A^0 $(= 10^{-8}\,\text{cm})$. Obviously, conventional optical microscopy will fail to resolve at the atomic scale.

Let us next consider an estimate of the wavelength of an electron. By the concept of wave-particle duality, waves may behave as particles and vice-versa. This is a consequence of quantum mechanics. For example, radiation or light waves display wave features which are characterised by their frequency, wavelengths etc. This is the classical behaviour.

Quantum mechanically, light waves will behave as particles with definite mass, spin, angular momentum etc. These are called photons which have zero mass and spin one. Likewise, classically electrons behave as particles with spin half and a very small mass of 9.1×10^{-31} Kg. Quantum mechanically, these electrons behave as waves.

The wavelength λ of any material particle is computed by Louis de Broglie's formula,

$$\lambda = \mathrm{h/mv}$$

where $\mathrm{h} = 6.626 \times 10^{-34}$ Joule sec is called the Planck constant, m is the mass and v, the velocity of the particle. The movement of an electron is effected by a potential difference. The energy acquired by an electron in moving across a potential difference of one volt is called an electron volt, eV. It equals,

$$1\ \mathrm{eV} = 1.6 \times 10^{-19}\ \mathrm{C} \times 1\ \mathrm{Volt} = 1.6 \times 10^{-19}\ \mathrm{Joules}$$

where e is the charge of an electron, given by 1.6×10^{-19} Coulomb.

If the electron moves across a potential difference of V volts, then the energy acquired is eV Joules which equals the kinetic energy,

$$\mathrm{eV} = 1/2\ \mathrm{m\ v^2} \rightarrow \mathrm{v} = \surd(2\ \mathrm{eV/m})$$

Putting this in de Broglie's formula, we get,

$$\lambda = \mathrm{h}/\surd(2\ \mathrm{emV})$$

Inserting the various numerical values for h, e and m, we obtain,

$$\begin{aligned}\lambda &= 6.626 \times 10^{-34}/\surd(2 \times 1.6 \times 10^{-19} \times 9.1 \times 10^{-31}\ \mathrm{V})\\ &= 12.25 \times 10^{-10}/\surd\mathrm{V}\ \mathrm{m}\end{aligned}$$

As a specific example if the electron moves through 10 keV, then $\mathrm{V} = 10^4$ and,

$$\lambda = 12.25 \times 10^{-12}\ \mathrm{m} = 12.25 \times 10^{-10}\ \mathrm{cm}$$

is the wavelength associated with that electron. Then,

$$\lambda_{\mathrm{light}}/\lambda_{10\ \mathrm{keV}} \approx 4 \times 10^4$$

showing that the resolving ability of the electron (at 10 keV) microscopy is 4 orders of magnitude superior than the optical one.

If the accelerating potential is 200 keV then,

$$\lambda_{200\,\mathrm{keV}} = 0.0025\ \mathrm{nm}$$

and,

$$\lambda_{\mathrm{light}}/\lambda_{200\,\mathrm{keV}} = 2 \times 10^5$$

which shows an improvement of 5 orders of magnitude. For higher energies relativistic effects become important and de Broglie's relation has to be suitably modified. This is beyond the scope of the present work.

In any experiment dealing with microscopes, focussing is an important issue. For optical microscopy a convex lens focusses all rays parallel to the principal axis at a single point, called the focus or the focal point. The distance of this focus from the center of the lens is the focal length (F). It depends on the curvature of the lens surface. This is, however, an idealised situation, true only for a perfect lens. In actual situation there are different kinds of aberration and the rays do not converge to a single point.

A perfect lens and different types of aberration are given below:

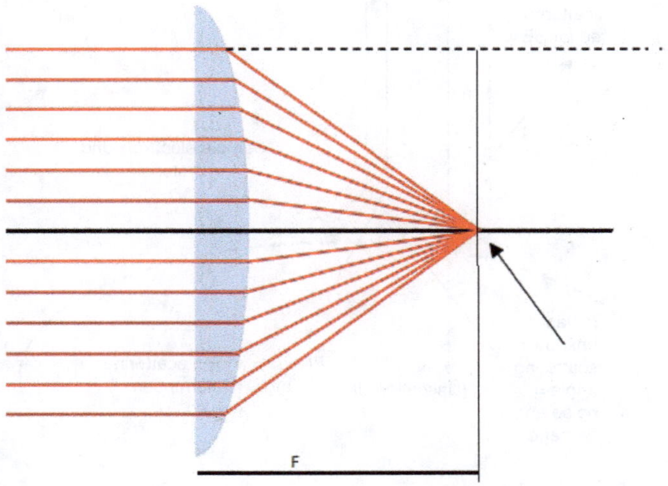

Figure 1: Rays are focussed at a single point called the focus.

Spherical aberration occurs due to a lack of uniformity of the incident angle at the surface.

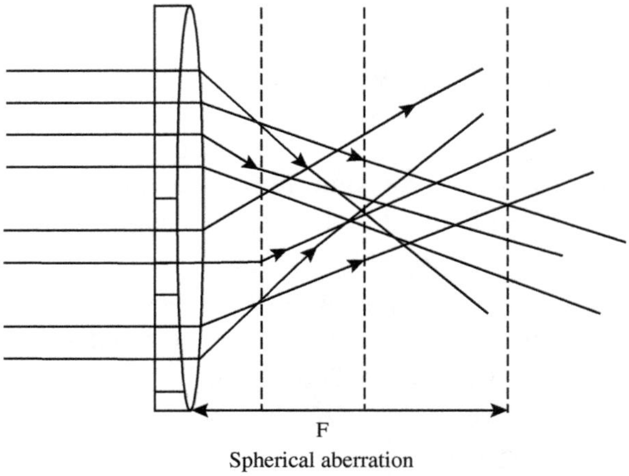

Figure 2: Spherical aberration.

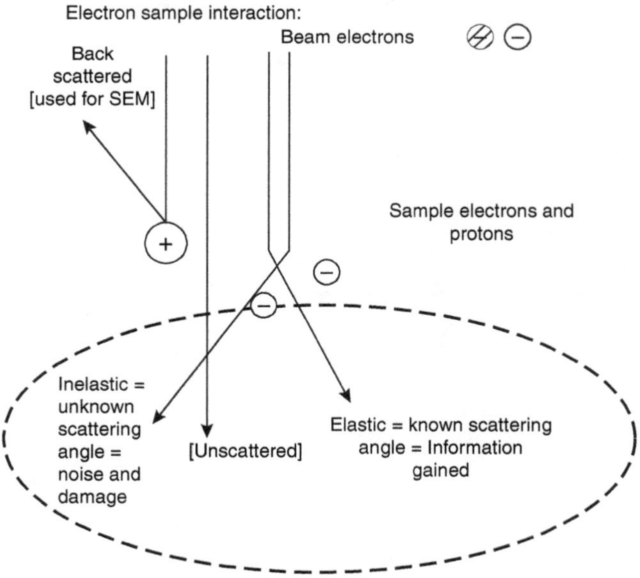

Figure 3: Electron sample interaction.

Chromatic aberration happens since light rays may not be monochromatic and are bent differently for different colors. Astigmatism happens when a lens has different focal lengths in X and Y planes. This will convert, for instance, a power spectrum into an oval shape instead of circular. In the microscope it is removed by stigmators.

In case of an electron microscope, there are many differences from an optical microscope. Focussing in an electron microscope is usually achieved by a solenoid coil of wire wound around a magnet. Passing a current through this coil generates a force that pulls the magnet inwards, thereby bending the path of electrons. By changing the current, the force and hence the bending can be altered and with it, the focal length. For an optical system, the focal length is fixed.

Light rays are composed of photons which are massless and charge neutral particles. These can easily pass through the atmosphere without any appreciable loss of energy or any bending due to scattering. Electrons, on the other hand, are negatively charged and possess a mass. Hence, to acquire maximal efficiency, the entire system (electron microscope) has to be operated in vacuum.

Vacuum system: It is organised in 3 areas — the gun, the column and the projection chamber, separated by differential apertures. The specimen area is pumped to 10^{-7} Torr, the gun area to 10^{-9} Torr.

Schematically we have,

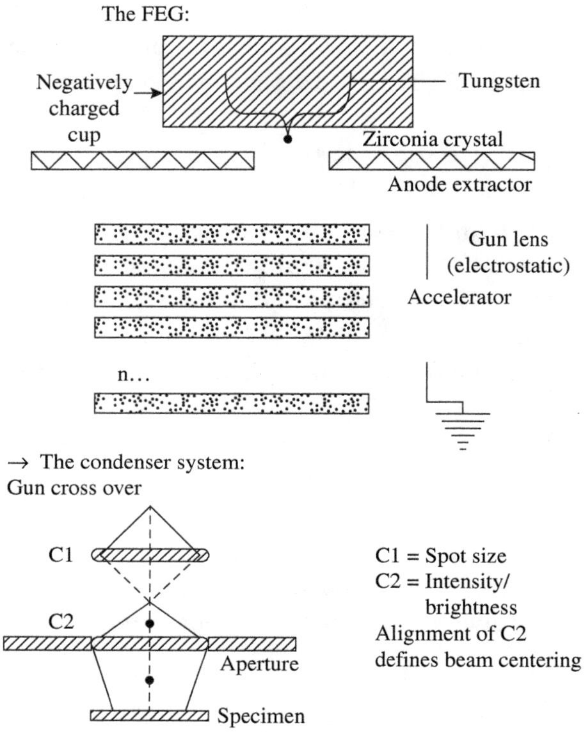

Figure 4: Working of the FEG (Field Emission Gun).

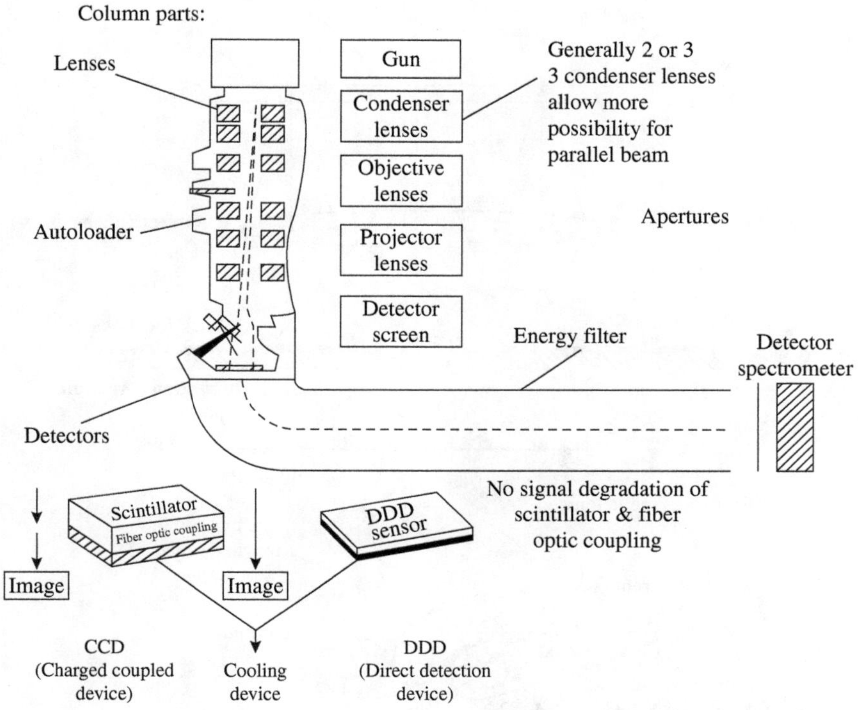

Figure 5: Working of FEG (Continued).

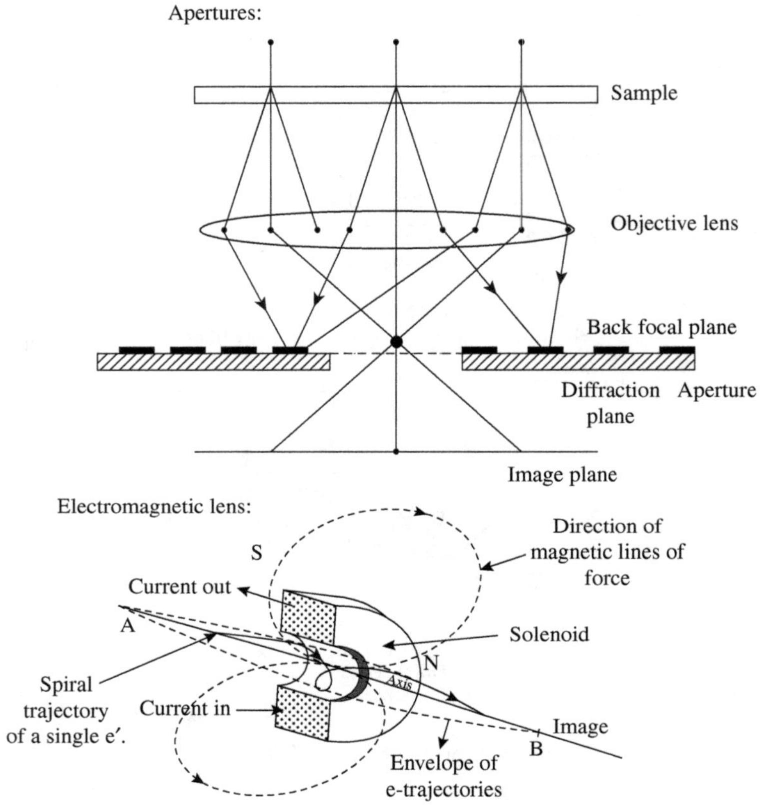

Figure 6: The elctromagnetic lens.

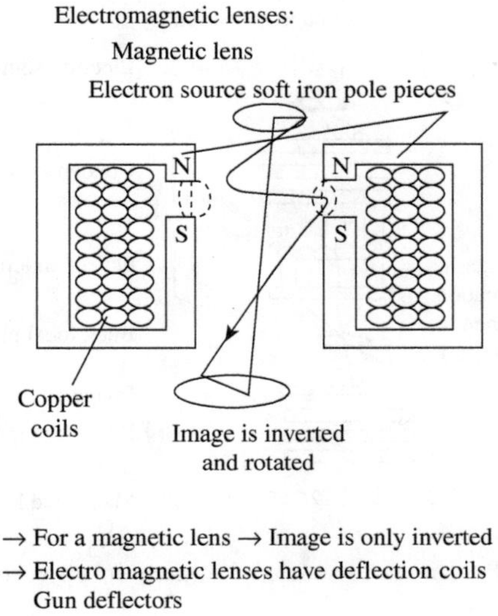

Figure 7: The electromagnetic lens (Continued).

Electron source: This can be tungsten filament or lanthanum hexaborate (LaBr) crystal or Thermionic source like field emission gun (FEG) (most important). The (FEG) works on the principle of applying a strong electric field on the surface of a metal that overcomes the binding energy of electrons causing them to move out of the metal surface.

Electrons may be inelastically scattered during electron-sample interaction. It leads to noise and damage. On the other hand, elastic scattering leads to information.

Image formation is of fundamental importance in any form of microscopy. First, we review the schematic of image formation.

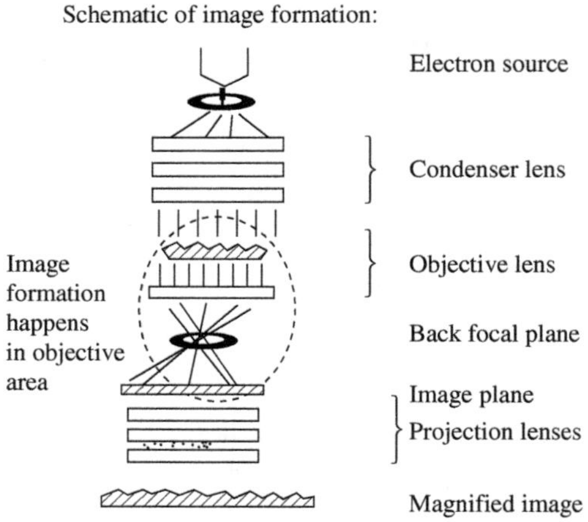

Figure 8: Image formation happens in objective area.

There are two distinct issues that play a pivotal role in image formation. These are called amplitude and phase contrast:

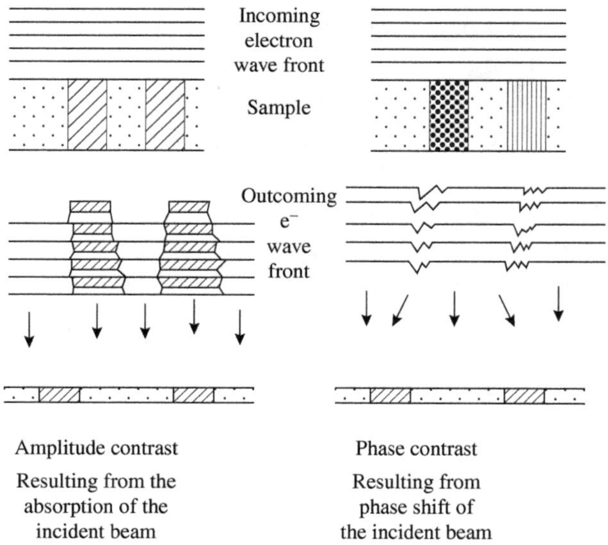

Figure 9: Amplitude and phase contrast.

The amplitude cantrast in cryo-EM biological sample is minimal (10%). The contrast is achieved by "transforming" phase shift into amplitude contrast.

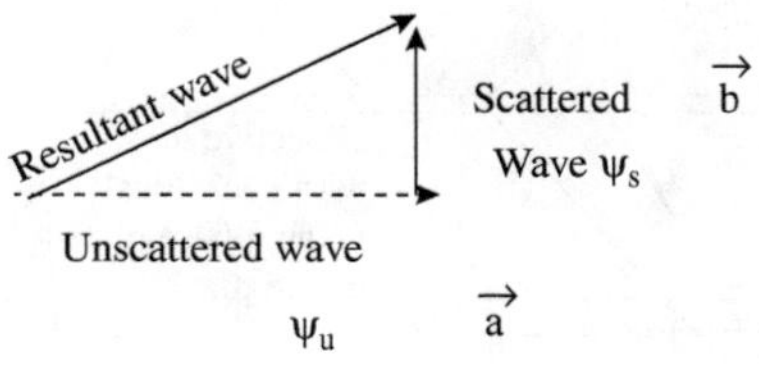

Figure 10: Increasing the contrast.

Here the resultant wave is given by,

$$\Psi_{\text{sam}} = \Psi_{\text{u}} + \text{i}\Psi_{\text{s}} = \vec{a} + \text{i}\vec{b}$$

- The image is formed by the interference between the unscattered beam and the scattered one. The modules of the 2 vectors are not very different.

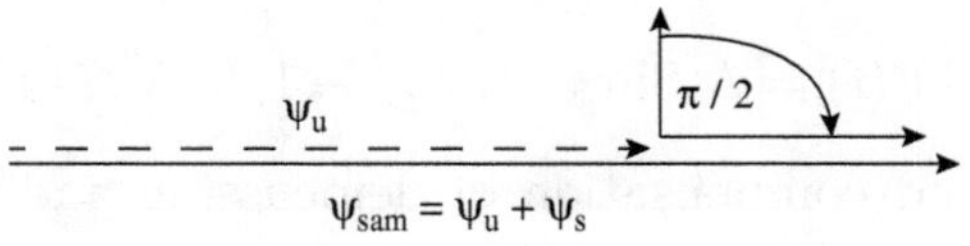

Figure 11: Shifting the phase of scattered wave.

- We want to shift the phase of the scattered wave by 90° to raise/lower the module of the resultant wave. This can be achieved by *phase plate*.

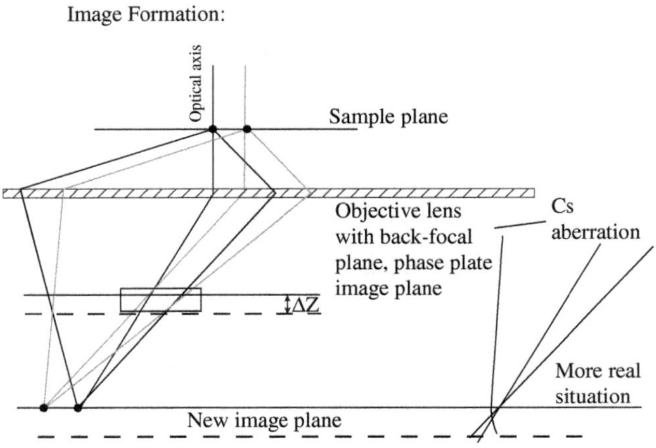

Figure 12: Amplitude modulation by aberrations.

Amplitude modulation by aberrations:

The modulation of amplitudes of the scattered wave by lens aberrations is easily described at the diffraction plane as a function of the scattering angle (i.e. of the spatial frequency that is represented).

$$\mathrm{CTF(f) = A\ Sin}\ (\pi\lambda\Delta \mathrm{zf}^2 - 1/2\pi\lambda^3\mathrm{Cs\ f}^4)$$

Measurement and compensation of defocussing and aberrations by Fourier processing of electron micrographs.

The image formed is affected by the CTF.

Electron-sample interaction damage:

- Electrons are damaging to biological samples.
- Need of focusing to be able to see on weak-phase scattering biological samples.
- Concept of low dose acquisition.
- Use of Energy filters.
- Direct Electron Detectors, motion correction and CTF.

The Pathologies and Horrors of cryo-EM images and how we can fix some of them.

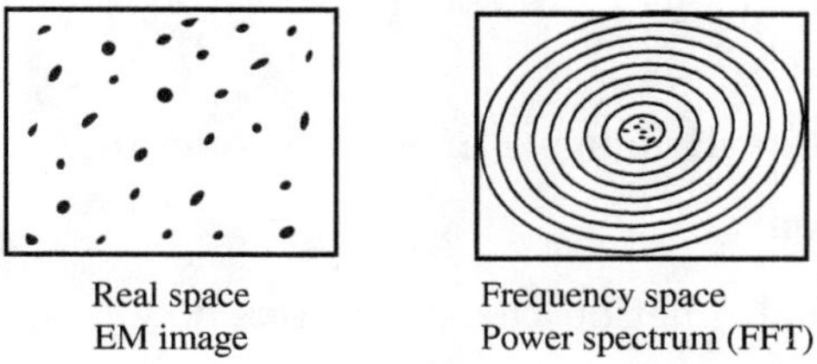

Figure 13: Fourier Space (in practise).

Fourier Space

Some details of Fourier space construction: consider 1D as a first example, $y = f(x)$.

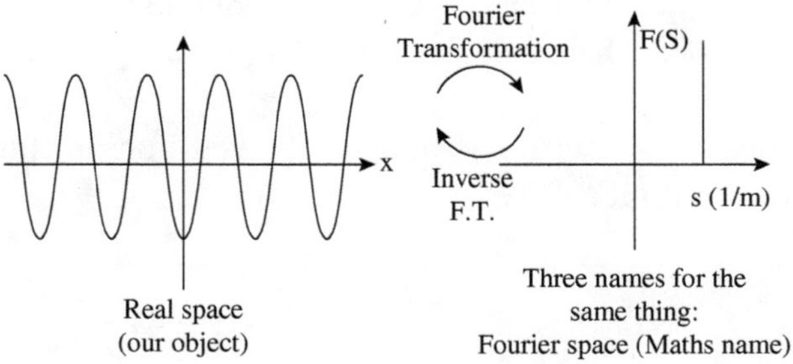

Figure 14: Properties of Fourier space.

It should be recalled that there are three names for the same algebraic relation dealing with Fourier space analysis:

- Fourier space (Mathematical name).
- Reciprocal space (Crystallography).
- Frequency space (Signal processing name).

Important outcome of Fourier Theory:

Fourier transformations (F.T.) occur in pairs — standard and inverse transformations. Note that the inverse of inverse F.T. reproduces the original F.T.

Fourier transforms work analogously in 2D and 3D.

Images and their Fourier Transforms

— So what happens if we remove some of the Fourire transform ("filter the image")?

The original (real) image starts to become blurry and lose focus.

Transfer Function:

- A methematical function that describes how a signal is changed.
- Well known example —

Transfer Functions in EM:

- Contrast Transfer Function (CTF) — describes the microscope. Multiplies the F.T. by an oscillating function depending on defocus and the spherical aberration coefficient of the objective lens.
- Detective quantum efficiency (DQE) — describes the camera. Expressed as a function of spatial frequency, it captures the performance of an X-ray detector to yield high signal to noise ratio (SNR) images.
- Defocussing messes with the images — even inverts the contrast.
- So why do we do it?

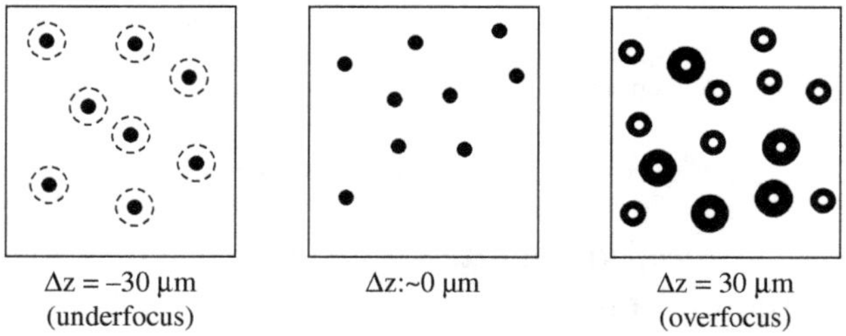

Figure 15: Contrast Transfer Function (CTF).

We typically need to defocus to see biological objects:

- How does the microscope change the levels? — and how can we change them back?
- Biological samples in cryo-EM are unstained and we need defocus to observe biological samples in cryo-EM. Defocus messes up with the images/inverts contrast.

The effect of the defocus on image contrast is described by the CTF.

$$\mathrm{F}(\mathrm{S_x}, \mathrm{S_y})\ \mathrm{F}(\mathrm{S_x}, \mathrm{S_y})\,{}^*\mathrm{CTF}(\mathrm{S_x}, \mathrm{S_y})$$

CTF at 0.3 μm.

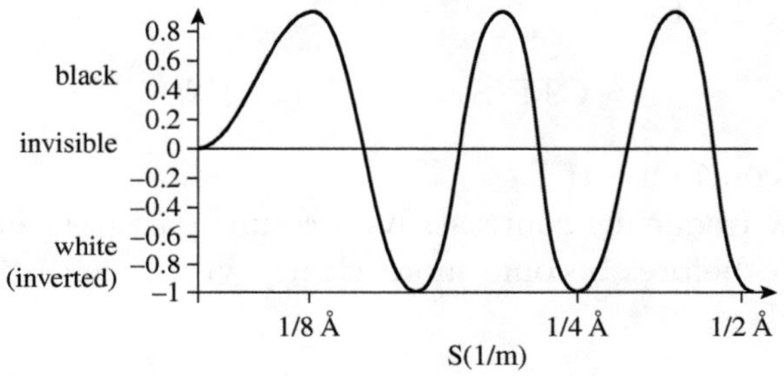

Figure 16: Power spectrum of EM image.

This is what you see when you look at the power spectrum of your EM image: the pattern of the CTF is

$$|\mathrm{F}(\mathrm{S})\,{}^*\mathrm{CTF}(\mathrm{S})|^2$$

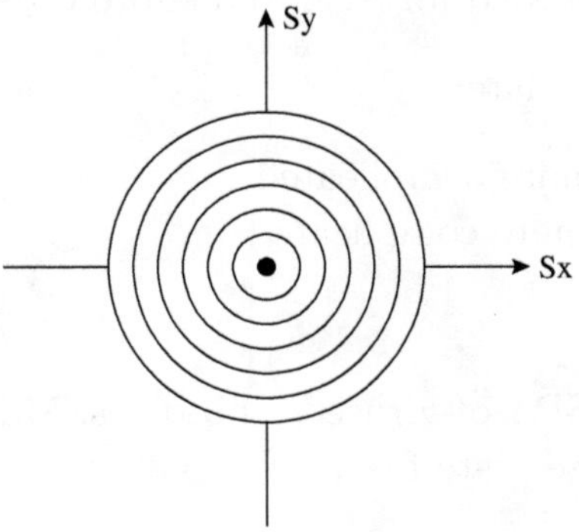

Figure 17: Power spectrum of EM image.

Features of some size get inverted contrast, others don't.
Low resolution features "disappear".

Higher defocus gives additional dampening of high frequencies. (spatial coherence envelope).

Cryo-EM with the phase plate:

- Introduces an additional phase shift of the unscattered vs scattered beam.
- Changes CTF to

$$\mathrm{CTF(S)} = \sin\left(\gamma(\mathrm{S}) + \Psi\right)$$

where Ψ is constant wrt S.

More low frequency contrast. As a result, particles that were not much visible before, become more clearly visible and distinct with the phase plate.

Advantage/Disadvantages of using a phase plate:

Why use a phase plate?

- Gives additional contrast that makes SPA posible on small proteins otherwise invisible in images.
- Improves contrast in (even low SNR) images of tomographic tilt series.
- Fewer particles needed for SPA and subtomogram averaging.

Why not use phase plate:

- Data collection more complicated.
- CTF correction more complicated.

Rules of thumbs:

- Don't use it for SPA of viruses, ribosomes, Mda complexes.
- Try without phase plate first.

CTF correction

Determining and correcting for the CTF is necessary for high resolution work in EM.

We need to dial back what the microscope has done to our image.

First steps of CTF correction is CTF determination:

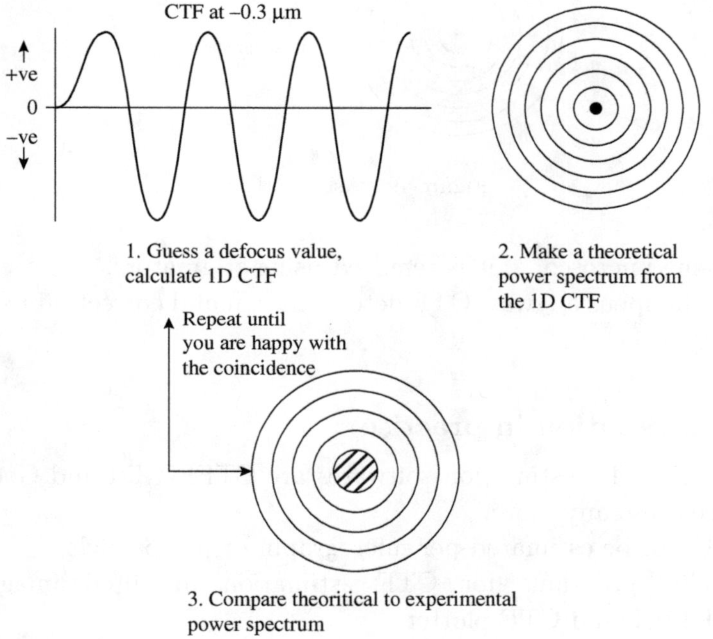

Figure 18: CTF correction/CTF determination.

Resulting outcome — An estimation of the defocus and astigmatism present in the image.

Side note: Astigmatism

When a lens has different focal lengths in X and Y (plane of the lens) it is called astigmatism. We see objective lens astigmatism as the power spectrum being oval (instead of circular). . . . we don't want an oval power sepctrum.

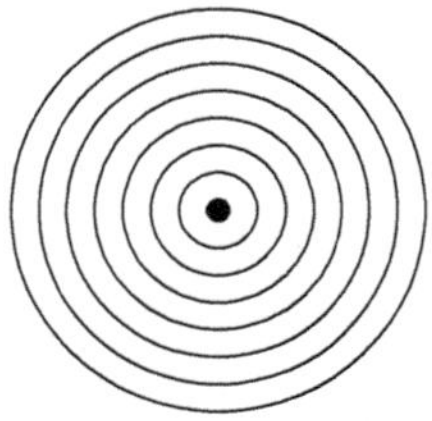

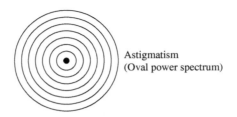

Figure 19: Astigmatism.

At the microscope, it is removed using stigmators.

If it happens anyway, CTF determination and correction may still work.

CTF estimation in practise

- Popular CTF estimation softwares are CTFFIND4 and Getf, but there are many more.
- CTF can be estimated per micrograph, or per particle.
- Specific programs for CTF estimation in tilted images — CTFTILT and CTF plotter.
- CTF estimation is often run through other software or ideally "on the fly" to estimate quality of images as they are recorded.

CTF correction:

Microscope has done:

$$\mathrm{F(S)F(S)\,{}^*CTF(S)}$$

Can we just divide by the CTF to undo that?

NO! (division by zero)

Some ways for correcting the CTF:

- Phase flip:

Simplest method: Change the sign in Fourier space in the areas where the CTF is negative. Used a lot in negative stain and tomography.

- Multiply by the CTF: Pragmatic solution since $\mathrm{F(S)}^*(\mathrm{CTF(S)})^2$ will have correct sign. This also supresses regions in Fourier space which contain mainly noise due to CTF ~ 0

- Wiener filter: Multiply Fourier Transform of image by $CTF(S)/((CTF(S))^2 + const)$ (this formula can be made more complicated to account for SNR etc.)

The "detectors" part of the presentation:

- Direct/indirect electron detectors.
- Electron counting (incl. Super resolution).
- Detective quantum efficiency (DQE).
- Drift/motion correction.
- Gain reference.

2) **Types of detectors:**

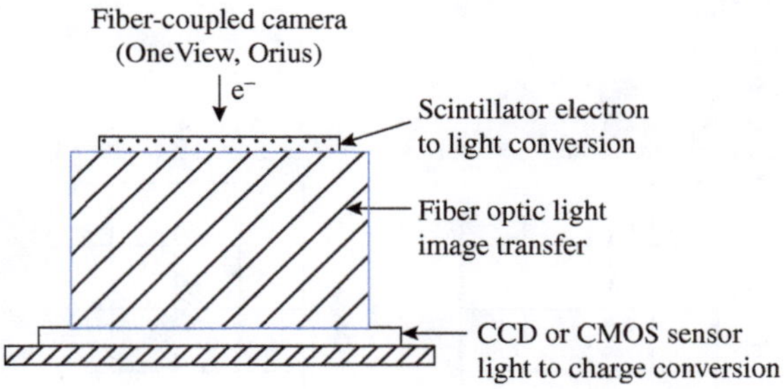

Figure 20: Types of detectors.

SciLifeLab microscopes have

- Direct detectors
- Gatan K2 summit (mounted on post column energy filter)
- Falcon 3EC.

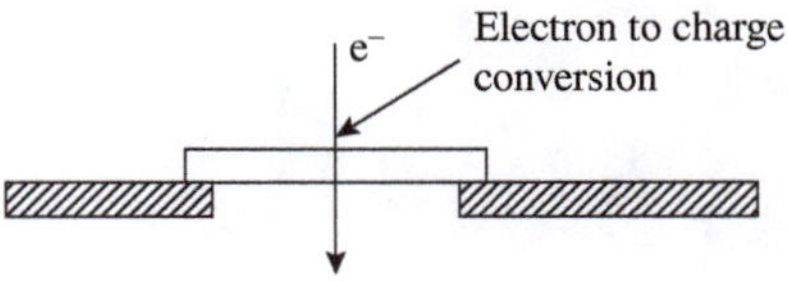

Figure 21: Direct Detection Camera (K2 summit, K2-1S).

Direct "detectors" "spread" the signal much less.

They are also more efficient at detecting nearly all electrons that impinge on the detector.

Advantages to direct detectors:

- Less spread/smear of the signal.
- Higher detection efficiency.

This in combination with faster readout also allows for:

- Counting of electrons.
- Motion correction.
- Integration or "Linear mode".
- Counting mode.
- Super resolution mode.

Integration or "linear" mode:

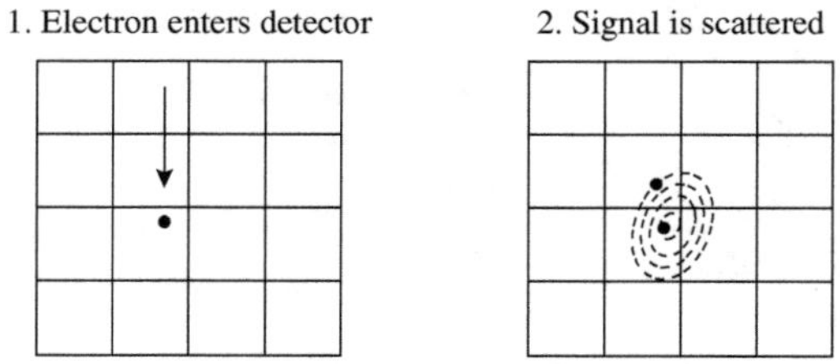

Figure 22: Electron enters detector.

1. Electron enters detector.
2. Signal is scattered.
3. Charge collects in each pixel.

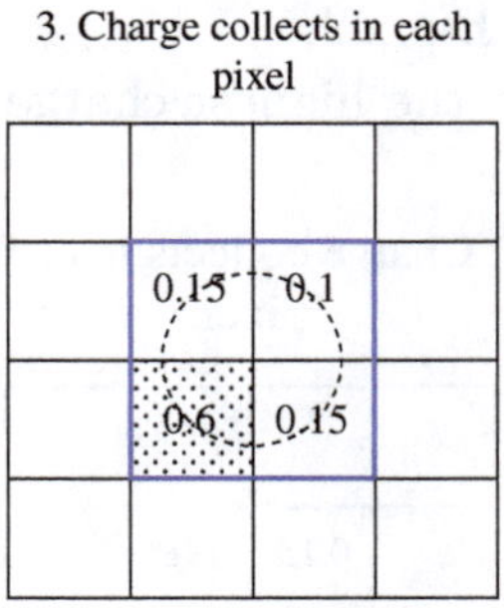

Figure 23: Charge collection in pixels.

4. Stop at step 3.
5. Charge integration.
6. Improved DQE at high frequency.

Counting mode:

1. Electron enters detector.
2. Signal is scattered.

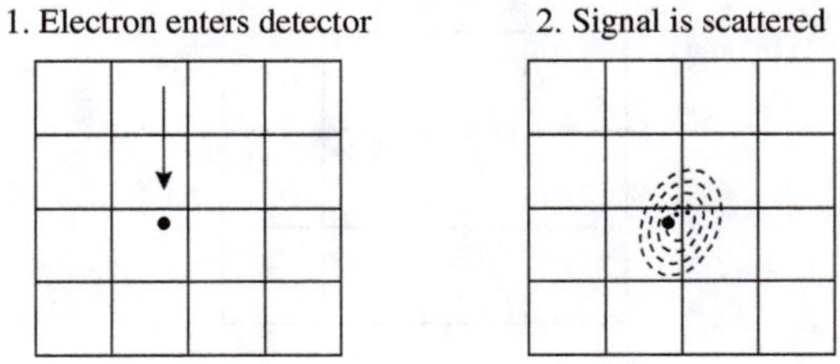

Figure 24: Counting mode.

3. Charge collects in each pixel.
4. Events are reduced to the highest charge pixels.

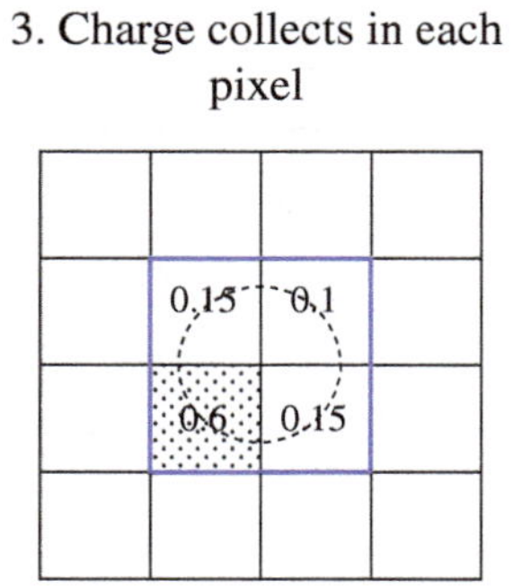

Figure 25: Charge collection in pixels.

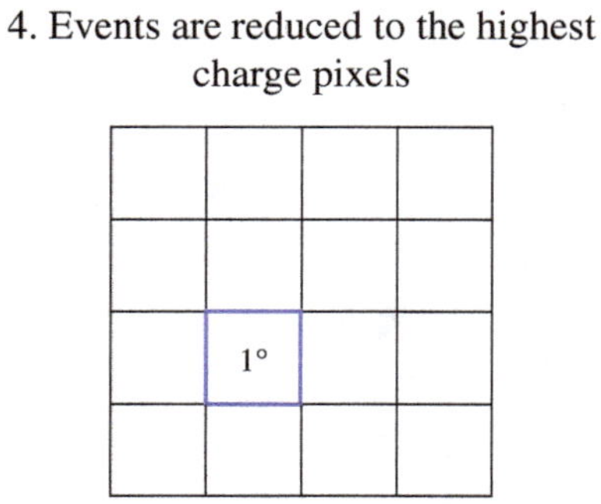

Figure 26: Reduction of events to highest charge pixels.

5. Stop at step 3. Charge integration improved DQE at high frequency.
6. Continue to step 4. Counting improved DQE at low and high frequency.

Super-Resolution mode:

1. Electron enters detector

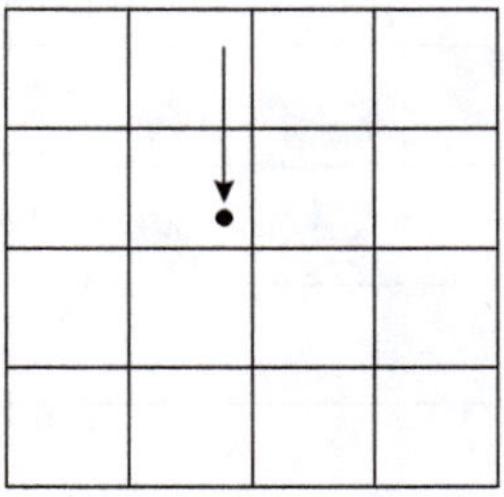

Figure 27: Super resolution mode.

2. Signal is scattered

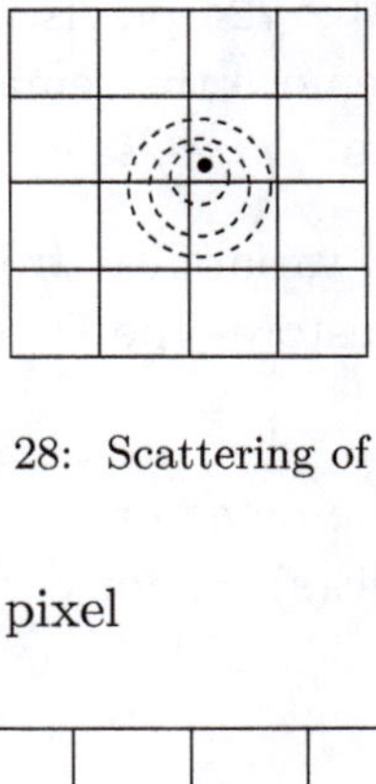

Figure 28: Scattering of signal.

3. Charge collects is each pixel

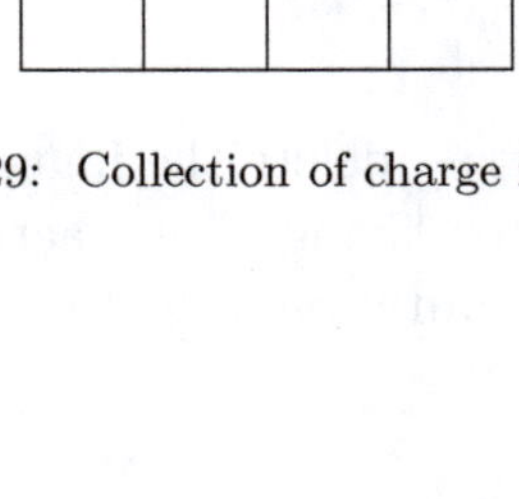

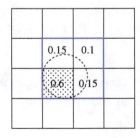

Figure 29: Collection of charge in pixels.

4. Events are localised with sub-pixel accuracy

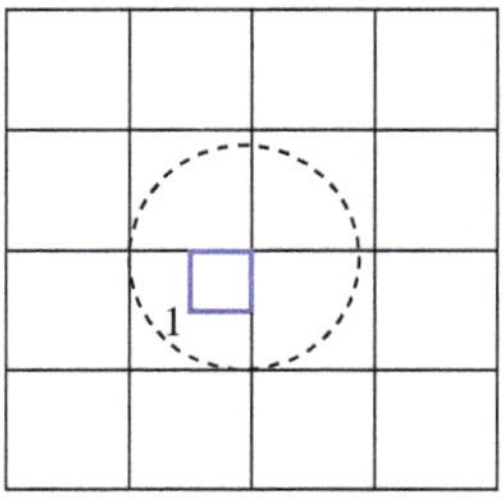

Figure 30: Localisation of events with sub-pixel accuracy.

5. Stop at step 3: Charge integration improved DQE at high frequency.
6. Continue to step 4: Super resolution improved DQE at low and high frequency and 7680×724 pixels.
7. Electron counting — Remove the remaining noise.
8. Typical dose rate of 10 $e^{-}/pixel/sec$

Counting removes the variability from scattering, rejects the electronic read noise and restores the Detective Quantum Efficiency (DQE).

- "Transfer function" of the detector.
- Perfect detector has DQE(s) =1 for entire frequency range.

Definition of DQE:

$$\mathrm{DQE(s)} = (\mathrm{SNR_{out}(s)/SNR_{in}(s)})^2$$

DRIFT:

Sample movement in one direction during exposure leads to blurring of high frequency details in that direction.

Drift/motion correction:

Classical solution to drift — delete the image

Direct electron detector "images" are actually movies with many frames. This enables computational drift correction, post exposure.

Simple cases of linear drift are easily corrected but things are not always easy and straightforward even at −196°C.

We can deal with more complicated motion in two ways:

- Per particle motion correction (Computationally expensive).
- Patch based motion correction, as implemented in Motion Cor2.
- Motion Cor2 with patch tracking.

Gain correction:

Fixing one final unwanted thing that the microscope does to our images.

- Pixels in any detector have slightly varying sensitivity.
- Fixing this is called gain correction.
- Take an image without a sample, with even illumination of each pixel.
- Invert the values of each pixel in the image and normalise it. This is the gain reference image.
- Correct images by multiplying them by the gain reference.

Fourier space in 2D and 3D — does that work?

	2D	3D
Real space	$I(x, y)$	$\rho(x, y, z)$
What is this in EM?	An image	A structure (a.k.a "map")
Function value is	Pixel intensity	The shielded Coulomb potential
Fourier transform interpretation	We look at $\|F(s_x, s_y)\|^2$ the power spectrum, sometimes called EFT	Sometimes called molecular transform.

Overview:

It is desirable to give a final overview emphasising the various points in cryo-EM.

Cryo-EM: Frome niche method towards new era in structural biology.

General overview of single particle analysis and 3D reconstruction.

How is the EM image formed?

- Thin specimen scatters electrons.
- Interference between scattered and unscattered electrons give phase contrast image.
- Image is 2D projection of original 3D object.
- 3D structure can be determined from a set of views at different orientation.
- Beam damage is the ultimate limit on resolution.

Projections and Sections:

Tomographic reconstruction

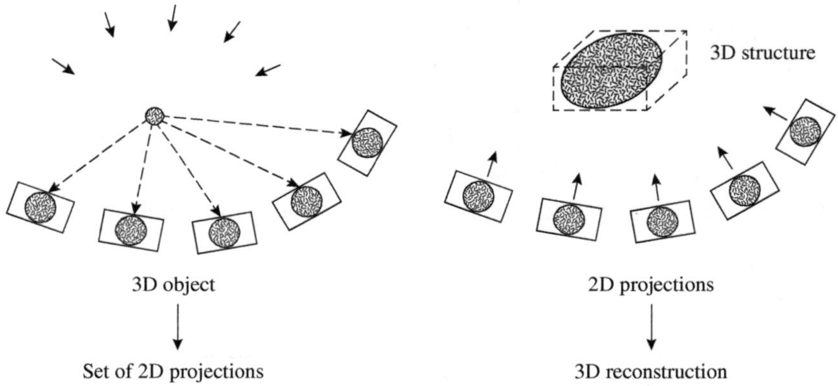

Figure 31: 3D reconstruction from 2D projections (Bulk-boundary construction).

The idea of reconstructing results in 3D from 2D is ubiquitous and quite powerful. It has been used in various fields, especially mathematical physics and related areas. Known as the 'holographic principle' it yields unexpected and significant insights. It has found applications in structural biology, although practitioners of different fields are not always aware of these connections since it is an area of modern research. Simply stated, large number of 2D projection has to be examined and the actual 3D structure, reconstructed. Larger the number of projections, better is the final reconstruction. In this context data processing plays an important role to finally determine accurately the 3D structure.

Single particle EM:

- Isolated macromolecular complexes.
- Randomly oriented in solution.
- No crystallization or ordered assembly needed.
- Position and orientation of each particle must be determined for 3D reconstruction.
- The more (homogeneous) particles used, the higher the resolution.
- Mixed states can sometimes be separated ("in silico purification").
- Interpretation by atomic structure docking or direct model building.

Negative staining	Cryo-EM
Simple procedure	More complex procedure
Quick to check samples	Longer time for checking samples
High contrast	Low contrast
Dehydration	Native, hydrated state
Possible distortion flattening	Near physiological conditions
	3D structure preserved

- Ideal image has high signal to noise ratio.
- Usually a cryo-EM image has low signal to noise ratio.
- Averaging similar views improves the signal to noise ratio. (S/N)
- Prior to averaging, alignment and classification are required.

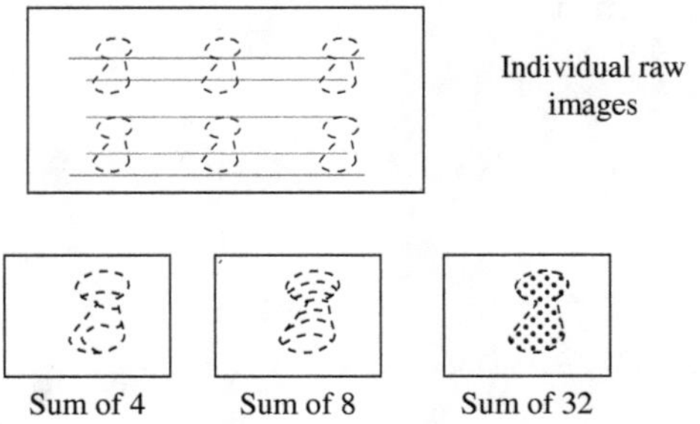

Figure 32: Alignment/classification of individual images.

Alignment strategies

- Reference free iterative alignment: 2 images are taken randomly and their average is used as reference for a 3rd image. This process is iteratively repeated until all particles are aligned.
- Translationally align to the rotationally averaged total sum of all images.
- Multi-reference alignment with classification in an iterative manner.
- Maximum likelihood approach does alignment and clasification together.

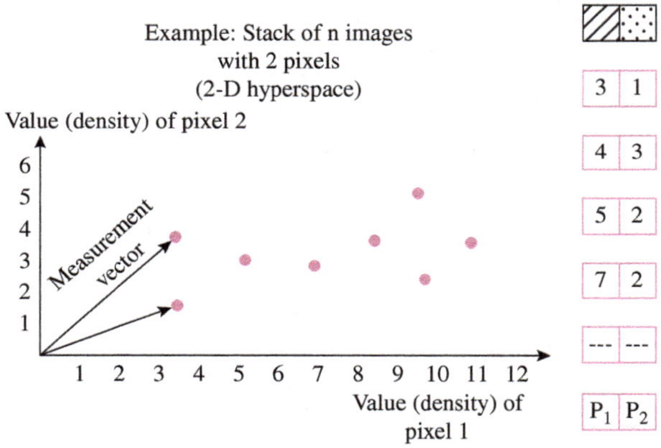

Figure 33: Classification of images: Multivariate statistical analysis.

Eigenimages can be drawn that represent the dataset and the main variance within it.

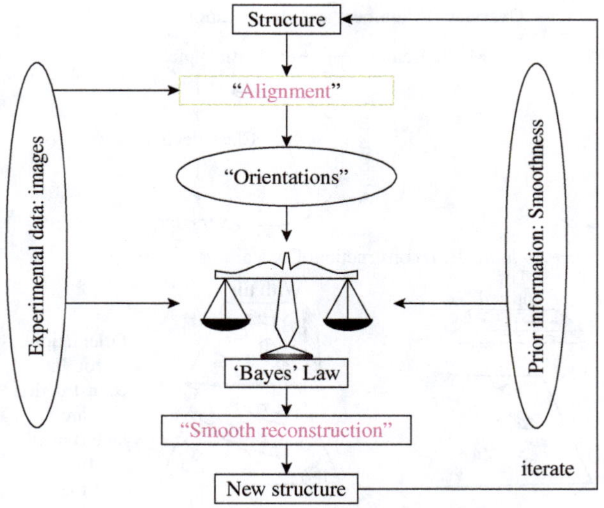

Figure 34: A probabilistic approach on cryo-EM determination.

Overview: Alignment and classification

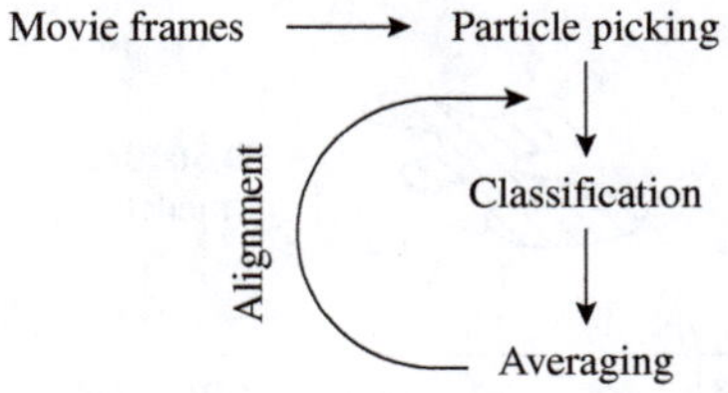

Figure 35: Overview of alignment and classification.

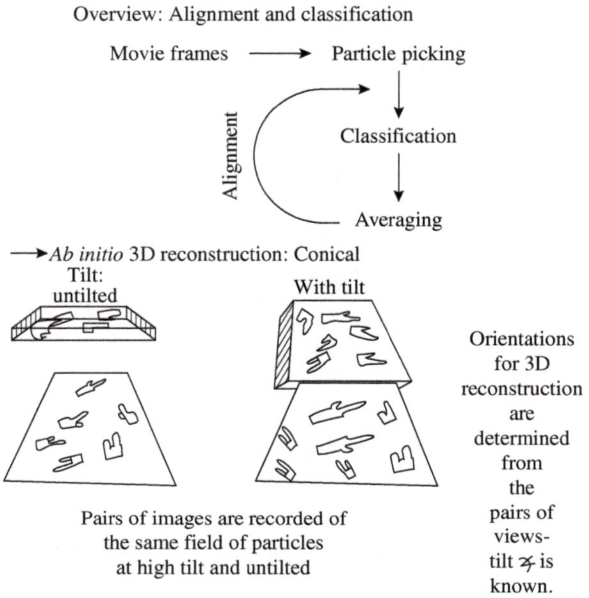

Figure 36: *Ab initio* 3D reconstruction: conical tilt.

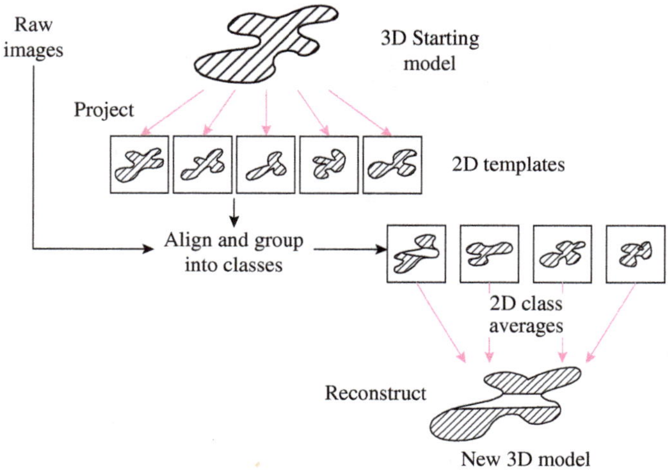

Figure 37: 3D reconstruction: projection matching.

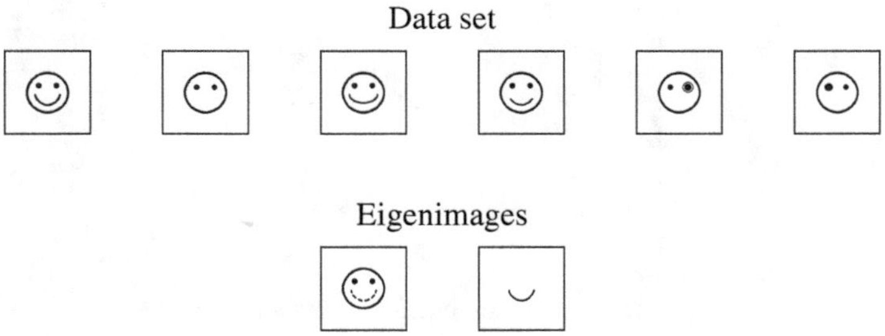

Figure 38: Eigenimages (principal component) analysis to detect and sort heterogeneous complexes.

It is useful to summarise various technological and software advances done in the context of cryo-EM analysis that have been discussed here.

- Coherent FEG sources and stable cryo images have provided big advances in resolution and throughput.
- Larger data sets and improved data processing lead to improved structures, including sorting of heterogeneous structures, merging of single particle and tomographic approaches.
- Direct electron detectors offer major gains in resolution and sensitivity movie mode enables drift correction restoring high resolution signal electron counting which enhances DQE at all spatial frequencies.
- Phase plates may significantly improve contrast and extend the limits of cryo-EM analysis of small proteins.

Alignment

- Determination of the relative positions and orientation of two images.
- Bring into register each image with a reference image.

To do so we need to find out the similarity between two images and we use the CCF (cross correlation function)

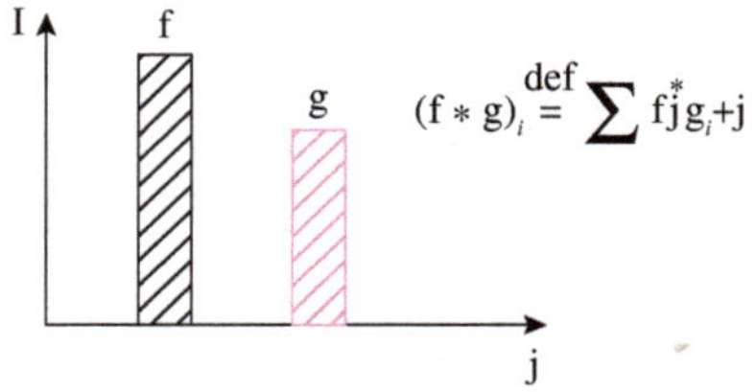

Figure 39: Cross correlation function (CCF).

The product is clearly maximised for the j translation so that when f and g are in register, all points contribute positively.

Alignment

An image B (central) is superimposed on top of the image of reference and for each rotation of B the CCF is calculated. The result is a curve with a peak on the maximum of correlation.

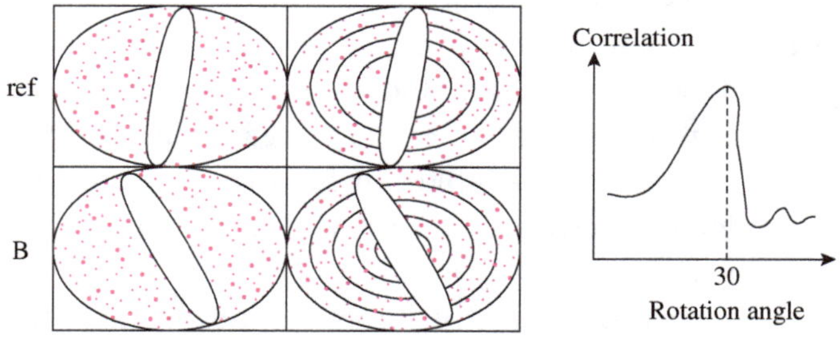

Figure 40: Correlation versus rotation angle.

Vitrification and how to work with cryo-EM samples:

- The human body consists of 60–75% water.
- All our cells and the cellular building blocks, DNA, membrane and proteins are in a hydrated solution.
- Immersion freezing for cryo electron microscopy (cryo-EM): Fundamentals.
- The high vacuum required in a EM greatly hampers the ability to study specimens naturally occuring in an aqueous phase.

Exposing "wet" specimens to a pressure significantly lower than the vapour pressure of water will lead to the water phase boiling off rapidly in the column, with devastating consequences for the structure of the specimen. Preparation methods, to dry specimens before inspection with room temperature EM are often associated with artefacts limiting the significance of the results.

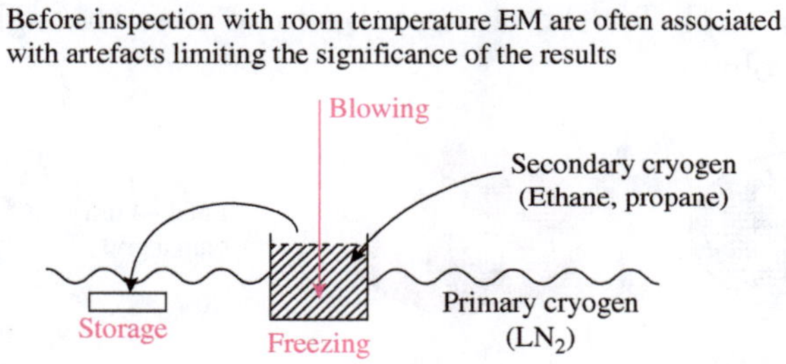

Figure 41: Working with cryo-EM samples.

Vitrification: Freezing of a hydrated solution without crystallization.

- Rapid freezing turns water into an amorphous solid state.
- Samples have to stay below $-140°C$.
- Vitrified samples are transparent for an electron beam.

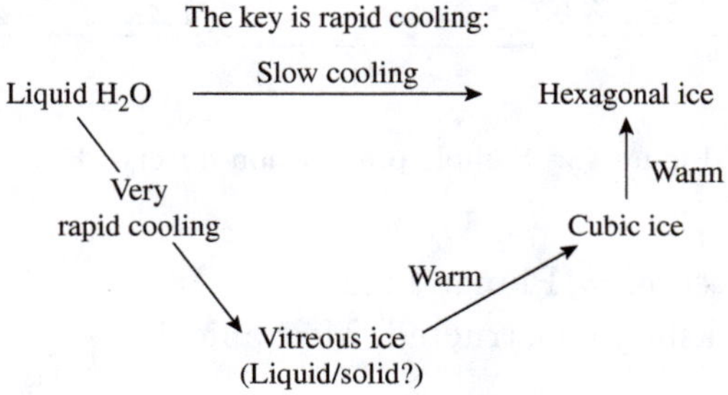

Figure 42: Effects of rapid/slow cooling.

- Rapid freezing = vitrification to −180°C in less than 20 ms
- Amorphous ice, like glass
- Transparent for electrons.

Sample preparation for cryo-EM:

- Use holey carbon film
- Create a thin film of sample in solution.
- Make it possible for the electron beam to pass through, minimize material volume.

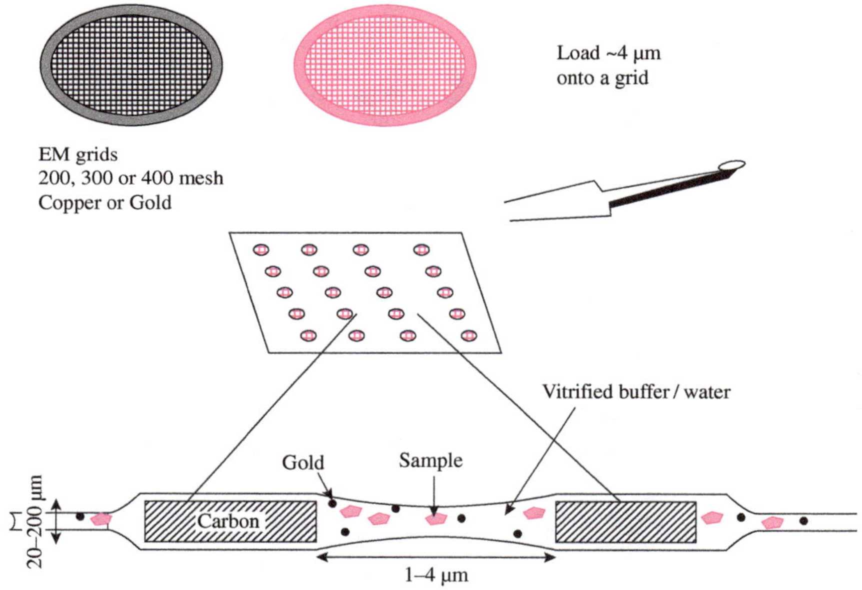

Figure 43: Sample preparation for cryo-EM.

- Immersion freezer = Plunge Freezer.
- Manual, workshop constructed, Mechanical

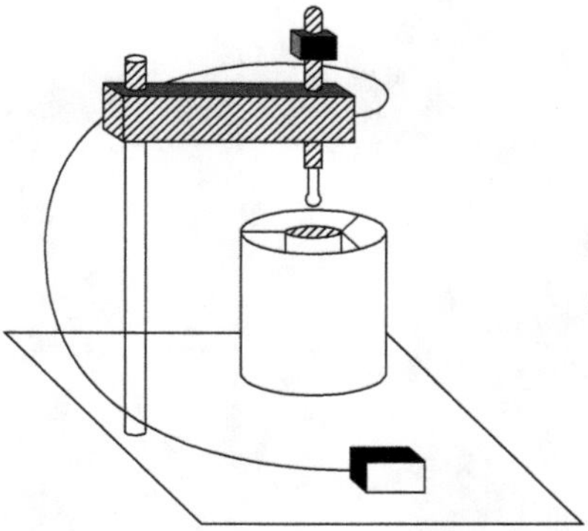

Figure 44: Experimental set-up.

Holey carbon (A)

- Quantifoil
- C-flat
- Lacey carbon (B)
- Holey gold film
- Dimensions of holes and support film (e.g. 2/2)

Add further support film:

Thin carbon layers or graphene
Glow discharge
Plasma cleaning in low pressure air
Remove dust
Possibilities to adjust the charges of the carbon surface

Chapter 3

Single Particle Cryo-Electron Microscopy

Cryo electron microscopy (cryo-EM) is a method applied on samples cooled to cryogenic temperatures. It is a high resolution technique which was developed in the early 1980s that uses an electron microscope with a cold stage and computers with very powerful image processing software to elucidate the 3D structures of biological macromolecules.[1] The technique of biological single-particle cryo-EM involves the flash freezing of proteins or other biomolecules and finally bombarding them with electrons to produce microscopic images of individual molecules. This information is used to reconstruct the 3D shape or the structure of the molecule.[2] These structures are fundamental in our understanding of how proteins work, what happens when they malfunction and how we can target them with drugs. Cryo-EM is generally suited to large, stable molecules which are easier to identify in low contrast images and can withstand electron bombardment without wiggling around too much; hence molecular machines which are made up of large protein complexes are often good targets.

---

[1]Yifan Cheng Single-Particle Cryo-EM at Crystallographic Resolution. Cell (Volume 161, Issue 3 p450–457 April 23, 2015); Brown, A., Long, F., Nicholls, R.A. ... Tools for macromolecular model building and refinement into electron cryo-microscopy reconstructions Acta Crystallogr. D Biol. Crystallogr. 2015; 71:136–153; Campbell, M.G., Cheng, A., Brilot, A.F. ... Movies of ice-embedded particles enhance resolution in electron cryo-microscopy, Structure, 2012; 20:1823–1828.

[2]https://www.nature.com/articles/aps2005169.pdf.

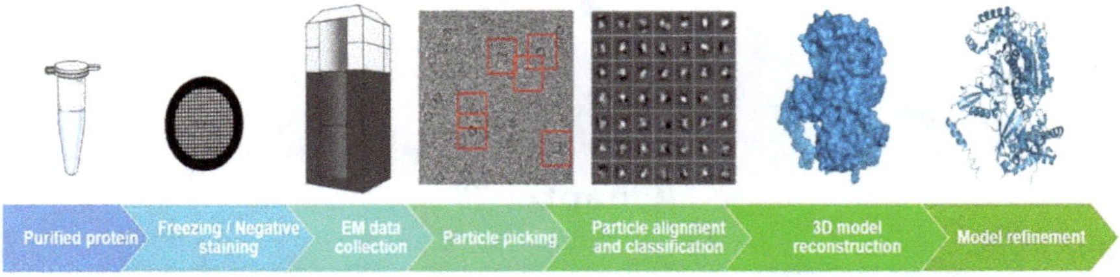

Figure 1: Cryo-EM workflow.

For decades, biochemists have been using X-ray crystallography: a technique of firing X-rays at crystallised proteins, to image biomolecular structures. But now, more and more labs are adopting the new cryo-EM method which allows users to take pictures of proteins that cannot be easily crystallised. Jacques Dubochet, Joachim Frank and Richard Henderson were awarded the Nobel Prize in Chemistry in 2017 for their joint work in developing the method. <The Nobel citation states that the prize was awarded "for developing cryo-electron microscopy for the high resolution structure determination of biomolecules in solution">

Cryo-electron tomography (cryo-ET) is a particular implementation of cryo electron microscopy where images of samples are shot at different angles by tilting them. This yields a chain of 2D images. The final image is generated by doing a 3D reconstruction. This is quite akin to a CT scan of the human body. In a cryo-ET study, biological samples are flash frozen, sliced in particular layers before imaging with an electron microscope. Multiple images are caught by tilting the sample along an axis. The final three dimensional picture or tomogram is found by suitably aligning these images and obtaining their resultant using computational techniques. In this chapter, I have used single particle cryo-EM as the main technique for structure determination.

Nyquist limit

The "Nyquist limit" is a common term used in single particle cryo-EM which refers to the maximum achievable resolution from a dataset. This term is derived from the "Nyquist-Shannon sampling theorem", which in the context of cryo-EM image processing refers to the maximum theoretical resolution imposed by the pixel size. The Nyquist limit refers to twice the pixel size; e.g. images recorded at 0.8 Å per pixel have a physical limit of achieving 1.6 Å resolution. In normal practice, achievable resolutions generally fall short of the Nyquist limit due to the dampening of high resolution signal from microscope and camera properties.[3]

Cryo-EM sample preparation

In this process an aqueous sample of a biological material (usually a purified protein complex) is applied to a support structure (grid) (see Fig. 2), thereby reducing its dimension to a layer that is as thin as possible (100–800 Å depending on the size of the biological molecule), and then freezing this thin layer fast enough to prevent the water

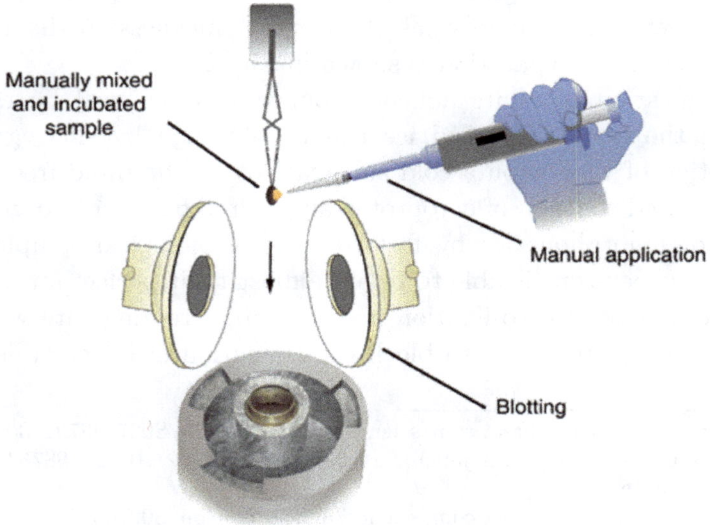

Figure 2: Traditional grid preparation.

[3]https://cryoem101.org.

from crystallising.[4] This method was first developed by Dubochet and colleagues in the 1980s. The method of sample preparation involves the following steps: (a) First, a fine layer of the suspension has to be built; (b) next, it has to be cooled into the vitreous state; (c) then shifting it to the microscope without rewarming higher than the de-vitrification temperature ($T_d = 140\,\mathrm{K}$) and lastly, (d) pondering it below T_d with an electron dose that is small enough to maintain the form of the specimen.[5] The construction of this thin layer of aqueous suspension may seem problematic at the beginning, due to the large surface tension of water, which maximises the volume to surface ratio. This obstruction can be bypassed when the desirable surface property is given to the supporting film.[6] A stable thin layer forms in the absence of a supporting film when the liquid is stretched over holes in a hydrophilic surface. In actual experiments, a drop of the solution is put on a clean, uncoated 200–600 μm mesh copper specimen supporting grid, most of the liquid is sucked with blotting paper and then the grid is vitrified after few moments when the ice in several grid squares has the correct thickness. A diagram of traditional grid preparation is shown in Fig. 2.

Vitrification of dilute aqueous solutions is achieved by rapidly cooling the liquid. Vitrified ice can be obtained by slow, gradual deposition of vapour on a cold substrate or by the rapid freezing of concentrated solution of cryoprotectant.[7] Cryo fixation into vitreous water or amorphous ice by fast freezing of biological samples like protein suspensions is able to retain almost their perfect structure.[8] This technique of cryo-fixation needs a rapid freezing rate which is fast enough (>100°C/s) to block the formation of ice crystals. The

[4]https://www.sciencedirect.com/science/article/abs/pii/S0076687916300295.

[5]https://academic.oup.com/jmicro/article-abstract/59/2/103/1988875?redirectedFrom=fulltext.

[6]https://www.sciencedirect.com/science/article/abs/pii/S0076687916300295.

[7]https://www.researchgate.net/publication/268875004_Principles_of_Cryopreservation_by_Vitrification#:~:text=Vitrification%20simplifies%20and%20frequently%20improves,populations%2C%20eliminates%20the%20need%20to.

[8]Pilhofer, Martin; Ladinsky, Mark S.; McDowall, Alasdair W.; Jensen, Grant J. (2010). Bacterial TEM. Methods in Cell Biology. Vol. 96. pp. 21–45. doi:10.1016/S0091-679X(10)96002-0. ISBN 9780123810076. ISSN 0091-679X. PMID 20869517.

consequence of this is that water droplets form a glassy, amorphous, metastable transient state.[9] The process of vitrification is able to maintain the structure of proteins and cells in almost their original states to atomic resolution. Vitrified samples are compatible with the vacuum conditions necessary for cryo electron microscopy.[10] A pragmatic method for cryo-preparation of biological samples is vitrification by plunge freezing in liquid ethane.

Grid screening and data collection

The ideal cryo-EM grid contains a rich distribution of monodisperse particles supported by a thin layer of vitreous ice. During the grid screening phase, images are recorded at a range of magnifications.

Grid view: A low magnification view of the grid gives information about the general ice quality of the sample.

Square view: Mid-range magnification views confirm ice quality and these images also reveal the presence of vitreous or crystalline ice, contamination and variation in ice thickness.

Hole view: High magnification views are required to evaluate the particle itself.

Image acquisition prior to direct electron detectors

Previously during the early days, electron micrographs of frozen, hydrated biological samples were photographed by either photographic film or employing scintillator-based digital cameras, similar to a charge coupled device (CCD) or complementary metal-oxide semiconductor (CMOS) (see Fig. 3).[11] The output from photographic

---

[9]Pilhofer, Martin; Ladinsky, Mark S.; McDowall, Alasdair W.; Jensen, Grant J. (2010). Bacterial TEM. Methods in Cell Biology. Vol. 96. pp. 21–45. doi:10.1016/S0091-679X(10)96002-0. ISBN 9780123810076. ISSN 0091-679X. PMID 20869517.

[10]https://www.researchgate.net/publication/268875004_Principles_of_Cryopreservation_by_Vitrification#:~:text=Vitrification%20simplifies%20and%20frequently%20improves,populations%2C%20eliminates%20the%20need%20to.

[11]https://link.springer.com/book/10.1007/978-3-319-49088-5.

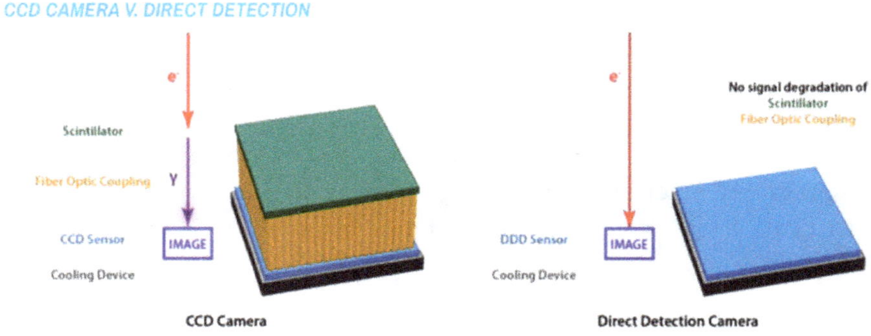

Figure 3: CCD camera versus direct detection camera.

films at high frequencies was normally superior than scintillator based cameras since it had a better detective quantum efficiency (DQE, which is an indicator of camera excellence as a function of spatial frequency).

Image acquisition using direct electron detectors

The large scale use of direct electron detection device (DDD) cameras ushered a 'resolution revolution' in single particle cryo-EM.[12] (See Mathematical Supplement to understand the particle nature of the electron as envisaged in these direct electron detectors vis-à-vis the wave nature of the electron revealed in electron microscopy through the use of Fourier analysis.) DDD cameras, with their rapid frame rate, leads to the capturing of cryo-EM data as dose fractionated image stacks rather than single micrographs.[13] This gathering of image stacks helps in fixing the stage or beam-induced motion thereby considerably enhancing the image quality by recovering the high-resolution signal spoilt by that motion. This technique improves image contrast and simultaneously eliminates from the final reconstruction large frequency noise that is caused by radiation damage.

---

[12]https://kids.frontiersin.org/articles/10.3389/frym.2023.1063909#:~:text=In%20this%20technique%2C%20electrons%20are,basic%20building%20blocks%20of%20life.

[13]https://pmc.ncbi.nlm.nih.gov/articles/PMC4409659/.

Cryo-EM data processing

The image processing pipeline implemented in cryo-EM data processing is needed to solve the 3D electron density of a target molecule, from noisy 2D images collected from Cryo-TEM (Transmission Electron Microscopy).

Each of these collected image is a movie of dose fractionated frames which need motion estimation and correction. These corrected micrographs are then used to estimate the microscope contrast transfer function (CTF), as well as to find and pick single particles. These single particles are extracted from micrographs and filtered and aligned using 2D classification methods. The final filtered particle stacks are used to perform *ab initio* 3D structure determination. These preliminary coarse structures are further classified and refined in 3D to get interpretable molecular density maps.[14]

A simplified cryo-EM data processing workflow is shown in Fig. 4.

Import movies

The analysis of cryo-EM images is significantly complicated due to their exceptionally low signal-to-noise ratio.[15] Direct detector device cameras (DDD) greatly increase the capacity of cryo-EM due to their improved detective quantum efficiency (DQE). These detectors use some kind of semiconductor technology that allows micrographs to be recorded as movies rather than individual exposures. DDD movies also revealed that one of the main source of image degradation was beam induced specimen movement. This also provides a way to partially correct the problem by aligning frames or regions of frames to account for this specimen movement.[16] The first step in the data processing pipeline is the import of one or more raw movie frames for processing.

[14]Single particle cryo-EM: dataprocessing techniques for obtaining optimal results, Ali Punjani, 2018 *Microscopy and Microanalysis*, *Volume 24*, *Supplement S1: Proceedings of Microscopy & Microanalysis 2018*, August 2018, pp. 2328–2329, https://doi.org/10.1017/S1431927618012126.

[15]https://pubmed.ncbi.nlm.nih.gov/32395065/ (Preprint Caryon).

[16]https://cryoedu.org chapter 1.

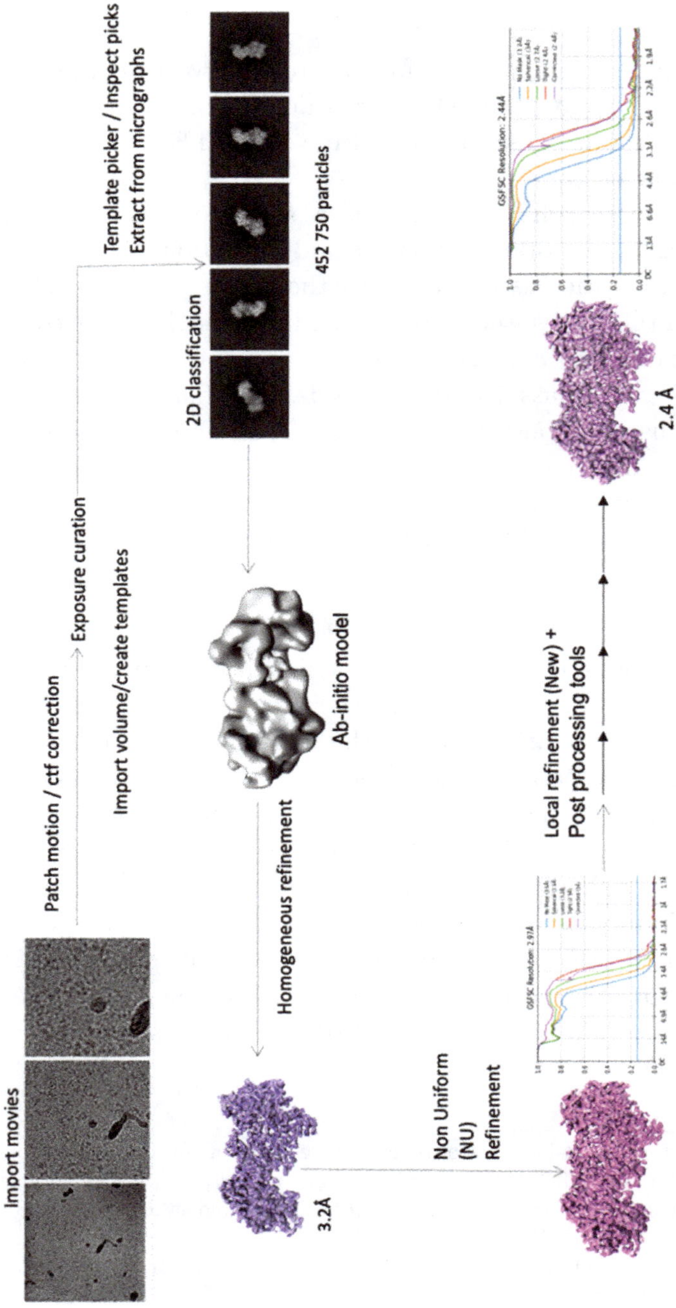

Figure 4: Simplified cryo-EM data processing workflow (LbNrdA sample).

Motion correction

During cryo-EM data collection, motion of the sample occurs for different reasons. The sample as a whole can move (stage drift), or sometimes the sample can deform due to the energy deposited by the electron beam.[17] This motion, as stated before, blurs the image and limits the resolution achievable by single particle cryo-EM.[18] Correction of this motion is the first step in the data processing timeline. The use of advanced DDD detectors enables data collection as dose fractionated movie stacks and implementation of the motion correction algorithm corrects for the error caused by beam induced sample drift and in turn restores the lost high-resolution information. As the electron beam induces doming of the thin vitreous ice layer, an algorithm was developed to correct for anisotropic image motion across the whole frame, which is suitable for both single particle and tomographic images. Patch-based iterative motion detection is combined with spatial and temporal constraints and dose weighting. MotionCor2: the multi GPU accelerated program is quite fast to keep up with automated data collection.[19] Patch motion correction corrects for both types of motion: due to stage drift as well as anisotropic motion and requires no previous knowledge of particle location. On the other hand, full frame motion correction only corrects for stage drift and treats the sample as rigid.[20]

CTF estimation

The method by which 3-dimensional electron density maps are abstracted from a collection of low dose, noisy cryo-EM images of separate macromolecules is called single particle reconstruction.[21] Cryo-EM sample preparation does not involve a fixation step unlike other

[17]https://myscope.training/CRYO_Motion_correction.
[18]https://pubmed.ncbi.nlm.nih.gov/28250466/.
[19]https://pubmed.ncbi.nlm.nih.gov/28250466/.
[20]https://myscope.training/CRYO_Motion_correction.
[21]https://pubmed.ncbi.nlm.nih.gov/28250466/.

microscopy techniques including other electron microscopy methods. The aim of this is to keep the sample in its native conformational state(s). We typically need to defocus to see unstained biological objects. However, defocus distorts the image and also inverts the contrast. The effect of the defocus on the image contrast is described by the contrast transfer function (CTF).[22]

In cryo-EM, a stream of electrons images samples of protein kept at ice-freezing temperatures. As this stream traverses through the sample it does not produce a big contrast in the image because generally the protein molecules and ice have similar transmission properties for electrons. This is overcome by operating the microscope in "phase contrast" mode. Then the self- interference of the electron beam, as it passes through the sample, generates contrast resulting in the imaging of variations in density. The electrons, on colliding with the protein molecules, are scattered. This scattering can be elastic or inelastic [In elastic scattering kinetic energy of the system is conserved. For example, if a photon is scattered elastically, there will be no change in its frequency (since frequency is directly proportional to the kinetic energy of the photon) but if it is scattered inelastically, there is a change in its frequency]. Only the elastically scattered electrons are relevant to the "phase contrast" image.

In quantum mechanics, electrons are characterised by a "wave function" which has an oscillatory behaviour in both space and time. The squared magnitude of this is the probability of finding the electron at a given position or moment when a measurement is made [For more details on this aspect see Mathematical Supplement]. When the cryo-EM sample scatters the electron elastically, the corresponding wave function also undergoes elastic scattering in several directions. The quantity of signal scattered at a particular angle is directly proportional to the spatial frequency of the sample. The various lenses of the microscope focus the scattered waves to construct an image. However, due to their varying scattering

[22]Erickson and Klug 1970 https://onlinelibrary.wiley.com/doi/10.1002/bbpc.19700741109.

angles, they travel different distances to reach the detector. The changes in path lengths are manifested as phase shifts in the electron wave function, which induces the scattered components of the wave-function to interfere among themselves. [The relation connecting path difference and phase shift is simple. A path difference of one wavelength corresponds to a phase difference of one complete cycle, i.e. 360 degrees or 2π radians]. This interference may be either constructive, destructive or inverting, which is governed by the spatial frequency. Consequently, the resulting image displays some frequencies more strongly than others, while some have their signs reversed. This phenomenon is characterised mathematically by the microscope's CTF.

The CTF can be described by the following equation:

$$\mathrm{CTF} = -\cos\left(\pi \Delta z \lambda_e f^2 - \frac{\pi}{2} C_s \lambda_e^3 f^4 + \phi\right)$$

where Δz = defocus, λ_e = wavelength of incident electrons, C_s = spherical aberration, f = spatial frequency, ϕ = phase shift factor.

Note that the argument of cosine appearing in the above definition is a function of defocus. Hence the comparative strength of different spatial frequencies is influenced by defocus, in the last image. Generally, a small value of defocus produces less contrast at small spatial frequencies. As cryo-EM images are very noisy, this indicates that particles at high defocus are easier to locate and pick out (both manually and algorithmically) than those at lesser defocus.

Effect of defocus on CTF

The most crucial and vital term in the CTF is based on defocus. For example, a possible way to defocus is to alter the power of the lenses in the microscope so that the sample gets out of focus. Also, if the sample is physically shifted from top to bottom and vise-versa along the microscope column, it moves the sample inwards/outwards from the focal plane, causing a similar effect as defocus. An important consequence of this is the fact that since cryo-EM samples are not flat, various portions of the sample will be at different levels of defocus. Since the CTF is related to defocus, a single image can have particles with varying defoci leading to different CTF values.[23]

Thon rings

Thon rings (see Fig. 5), also known as CTF rings, are a phenomenon observed in the power spectra of micrographs by bright-field (BF) transmission electron microscopy (TEM) imaging.[24] These rings are the effects of the contrast transfer function (CTF) which modulates the Fourier transform of the object in a defocus-dependent way.

The white rings correspond to the CTF maxima and the dark rings indicate spatial frequency bands that have no signal.[25]

CTF correction

Once the CTF has been estimated we need to carry out CTF correction. The aim of this correction is to restore the signal by

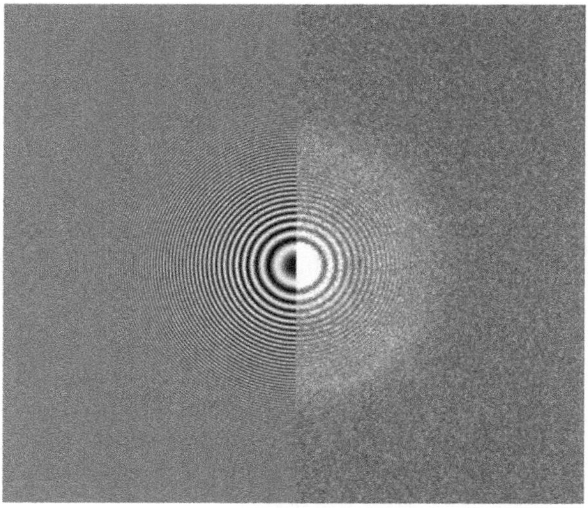

Figure 5: Power spectrum of a typical BF TEM image with concentric thon rings.

[23]https://pubmed.ncbi.nlm.nih.gov/28250466/.
[24]F. Thon, Z. Naturforsch 21 (a) (1966) 476–478.
[25]https://www.globalsino.com/EM/ Page 4236.

reversing the sign of the inverted signal. This is usually achieved by phase flipping, where the sign is simply inverted, by multiplication with the CTF or by using a Wiener filter[26] where the CTF is divided by a frequency-dependent weighting function.

Gain correction

We have to fix one final unwanted thing that the microscope does to our image. The raw images obtained can show detectable stripy patterns. This happens because pixels in a detector have slightly differing intensity and fixing this is called "gain correction". This is carried out by taking an image without a sample with even illumination of each pixel. The values of each pixel in the image are inverted and normalised. This is the gain reference image. The raw images are corrected by multiplying them with the gain reference.

Movies by Exposure

We can manually curate a set of exposures. This step requires classifying a set of micrographs or movies, using either population level statistics and controllable thresholds or through the separate observation of diagnostic plots. Here the set of micrographs or movies can be classified, depending on CTF fit, mean defocus, astigmatism, motion trajectories and power scores related to the collected particles. Similarly, the CTF diagram and thon rings, CTF fit, rigid and local motion evolution plots associated with a micrograph can also be examined. Based on this data, micrographs can be accepted or rejected for next level processing.

Particle picking

After the curation step, the most obvious next step is particle picking. In this step, we display the CTF corrected, curated micrographs and try to locate our particles based on size and shape. Once the particles

[26]Wiener, Norbert (1949). *Extrapolation, Interpolation, and Smoothing of Stationary Time Series.* New York: Wiley. ISBN 978-0-262-73005-1.

are found, we pick them manually. It is important to pick particles from micrographs having a range of defocus levels (both high and low defocus). Once a number of such different micrographs have been manually picked, the particles are extracted with the correct box size and used for 2D classification to make templates. These templates formed can then be used for template-based picking which can pick particles from all the micrographs. Another kind of picking that is gaining importance is the deep learning-based particle pickers. Since, manual picking can be a long and tedious process, different automatic particle pickers have been developed. One such particle picking software is crYOLO, which is based on the deep learning object detection system You Only Look Once (YOLO). The network is trained with 200–2500 particles per dataset, after which it can automatically recognise particles with high precision.[27]

Particle extraction

The next step after template based picking from all the micrographs is particle extraction. If we make the box size too small, there is a risk of cutting out signal. However, if the box size is too large, lack of RAM becomes an issue: this can be an important aspect in image processing since high resolution single particle structures are known to require hundreds of thousands of particles.

Some of the key factors for choosing the right box size:

Pixel box size affects the computational time. The optimal box sizes to use for image processing are those that are even numbers with a low prime factor such as 2, 3, 5 and 7.

CTF signal can be cut off with small box sizes: In the step of CTF estimation, particles can recuperate high resolution data and improve the final 3D reconstruction. However, if the extracted box size is too small, it can cut out the CTF signal and CTF correction will not be done properly, which in turn restricts the resolution. To

[27]https://www.nature.com/articles/s42003-019-0437-z.

avoid this, a box size of about 150 Å larger than the average diameter of the particle is often chosen.[28]

2D classification

The low signal-to-noise ratio (SNR) of cryo-EM images and the change that occurs due to CTF effects make it problematic to evaluate the images in a particle stack. The conglomeration of similar particle images, popularly termed as 2D classification, and computation of class average images is a good indicator regarding the contents of the dataset. Secondary structural characteristics like α-helices are frequently seen in 2D class average images from high-resolution data. This process of clustering similar particle images was first put forward by van Heel and Frank.[29] The principal objective of 2D classification is to make patterns appear from data. Averaging of similar views improves the signal-to-noise ratio. Before averaging, alignment and classification are required. Different alignment strategies can be used. Some of the most popular are the following:

Reference free iterative alignment: Two images are taken in random and their average is used as a reference for a third image. The process is repeated iteratively until all particles are aligned.

Maximum likelihood approach: This method does alignment and classification together.

The maximum likelihood method is implemented in Relion.[30] We commence with a set of arbitrary reference images. For each particle image, probabilities related to its rotation, translation and the quantum of matching to each reference are calculated. The 2D Fourier transforms of the translated and rotated particle images

---

[28]https://jiang.bio.purdue.edu/how-big-should-my-particle-box-be/#:~:text=To%20avoid%20this%20issue%2C%20use,260%20%C3%85%2C%20or%20325%20pixels.

[29]Frank J, van Heel M. Correspondence analysis of aligned images of biological particles. *J Mol Biol.* 1982, Oct 15; 161(1):134–7, https://doi.org/10.1016/0022-2836(82)90282-0. PMID: 7154073.ef.

[30]https://relion.readthedocs.io/en/release-5.0/.

are taken together and composed to form the overall image. These averaged images are chosen to be a new set of references, and this iteration is continued several dozen times. The final result is a set of representatively averaged images.[31]

2D classification and single particle reconstruction (SPR) are rather similar. For either case, each particle image is aligned and compared against a set of reference images after which the reference is updated. Their only distinction is that SPR includes the condition of consistency with a specific 3D structure which is missing for 2D classification.

An initial notion of heterogeneity in a dataset is obtained from 2D classification. Exploiting this method, we are often able to visualise patterns of particles in various configurations or differing sizes and separate the individual particle images by using the 2D class identities (see Fig. 6). These class average images also provide an overview of the range of directions available in a dataset, from which observations are feasible. Example: it is very simple to identify 'top' from 'side' views of symmetric particles and the symmetry

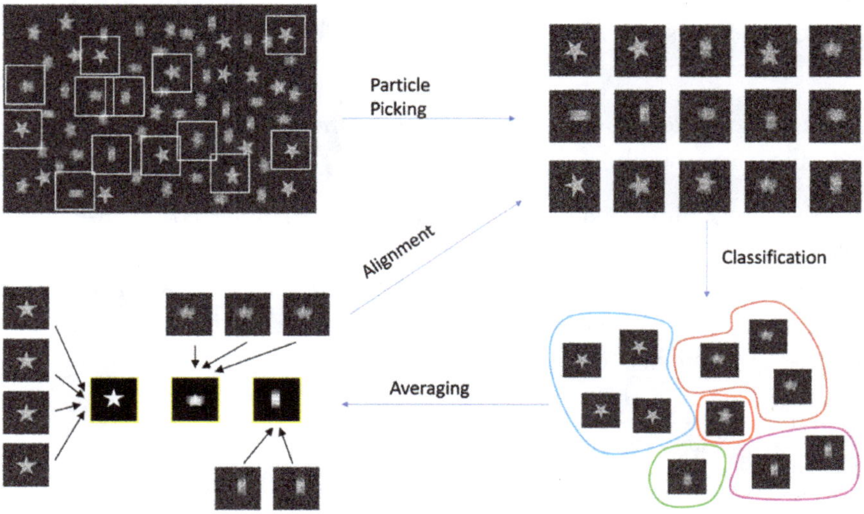

Figure 6: (Overview of alignment and classification; individual images from Marta, cryo-EM course).

[31]Methods in Enzymology, Vol 579, 2016 pg-125–157.

point group is occasionally recognisable.[32] 2D classification quickly sorts particles into multiple 2D sectors to enable stack cleaning and removal of unwanted (junk) particles. Also, it is extremely efficient for checking particle quality before proceeding to 3D reconstruction.[33]

Nearly all my experience with cryo-EM data processing has been obtained from cryoSPARC, so this chapter will focus on that program's algorithms.

Ab initio model generation

An algorithm was developed to make it possible for the first time to perform unsupervised 3D classification, where multiple 3D states of a protein can be identified from one unique protein without user input and also without the prior knowledge and also without the assumption that all 3D states resemble one another. Another algorithmic development was used to quickly accelerate high resolution refinement of cryo-EM maps. It exploits features of image alignment to attain great advantages in computations. This is achieved by eliminating redundant computation. They are installed in a separate package (cryoSPARC) (cryo EM Single Particle Ab-Initio Reconstruction and Classification) <cryoSPARC: Algorithms for rapid unsupervised cryo-EM structure determination>

In this step, we reconstruct a single (homogeneous) map or multiple (heterogeneous) 3D maps from a set of particles, without the need for initial models or starting structures. There are different methods for doing 3D reconstructions. The most common method is 3D reconstructions from 2D projections (see Figs. 7–9).

Another way of carrying out 3D reconstruction is through projection matching. We collect data in 2D but we need information in 3D. In single particle cryo-EM, many identical samples are imaged in different orientations.[34] One of the important things needed to get a high resolution structure is the ability to image the 3D object from as many orientations as possible so that we get views from all sides, top view, side view, tilted view etc. With this information in hand,

[32]https://pubmed.ncbi.nlm.nih.gov/26705325/.
[33]https://cryoem101.org/chapter-5/.
[34]https://pubmed.ncbi.nlm.nih.gov/26705325/.

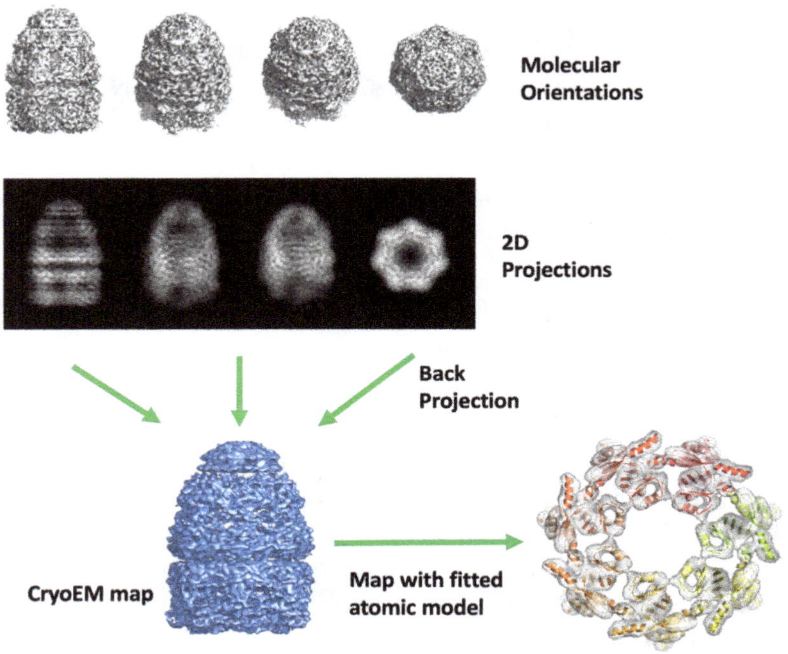

Figure 7: Figure remade: Individual images taken from Colin Palmer's cryo-EM slides, ccpEM.

it gets easier to reconstruct a reliable 3D volume, particularly when there are particles present in the dataset that can be grouped into all the different views possible.

We can imagine this if we think of shining a flashlight on a ball. To get a proper idea of the ball's structure, we should shine the flashlight from as many orientations as possible (see Fig. 8).

Heterogeneous refinement

Heterogeneous refinement is a method implemented in cryoSPARC which is very similar to 3D classification in Relion. It simultaneously classifies particles and refines structures from several initial structures, which are usually obtained following separate *ab initio* reconstructions. This process facilitates the observation of small differences between structures which might not be visible at low resolutions. It

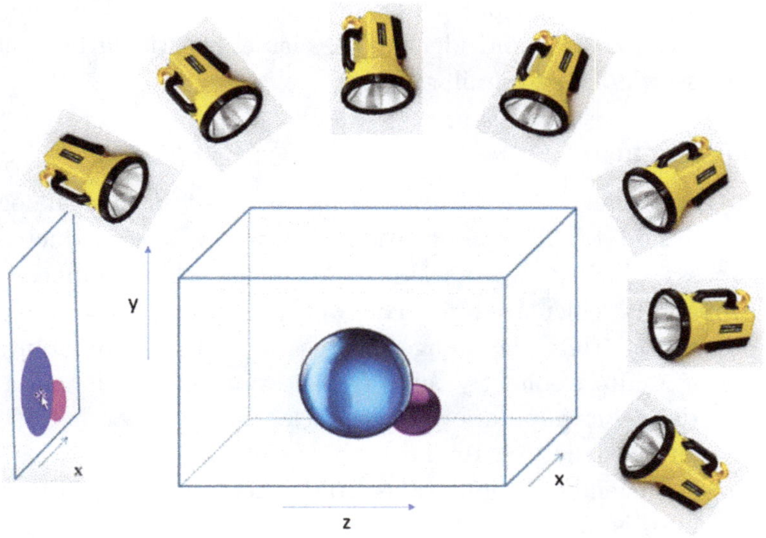

Figure 8: Flashlights shining on a ball. Increasing the number of flashlights helps in improving the 3D reconstruction from 2D images.

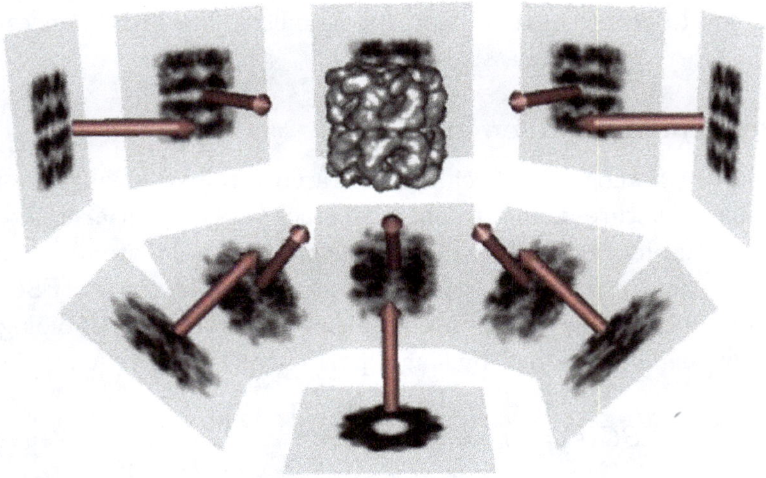

Figure 9: 3D reconstruction from 2D projection.

also helps to re-classify particles and aids in sorting. This method helps to sort particles and identify 3D classes: particularly in cases where classes look very similar.

3D classification

As stated previously, classification of 3D models is rather analogous to the method of 2D classification. Normally, it is inducted in the process of SPR by statistical weighting. This implies that the probability calculated for each particle image also contains the probability that the image arises from all the different 3D densities.[35] Quite frequently, successive sessions of classification are used to define a population of particle images that yields the best resolution structure. In short, the segregation of particles into separate homogeneous populations from a heterogeneous mixture is achieved by this 3D classification.

Reasons why we might need 3D classification when 2D classification could potentially sort out the heterogeneity is that heterogeneity may be less obvious at the 2D level and smaller differences cannot be sorted out in 2D classification. Also, some differences cannot be seen at 2D level because of their complex geometry. These differences can only be sorted by 3D classification.

Homogeneous refinement

The job of homogeneous refinement rapidly refines a single homogeneous structure using a particle stack as an initial reference to high resolution. It is validated using the gold standard FSC (Fourier shell correlation). <Jeol, Glossary of TEM terms> FSC is a measure of reliability of the three-dimensional structures of biological macromolecules obtained by single particle analysis (SPA).

$$FSC(k) = \frac{\sum_{k,\Delta k} F_1(k) \cdot F_2(k)^*}{\sqrt{\sum_{k,\Delta k} |F_1(k)|^2 \sum_{k,\Delta k} |F_2(k)|^2}}$$

[35]Fred Sigworth; Principles of cryo-EM single particle image processing, *Microscopy*, Volume 65, Issue 1, February 2016, Pages 57–67, https ... Abstract · Introduction · Theory of SPR · The image-processing pipeline.

where $F_1(k)$ and $F_2(k)$ are 3D Fourier transforms of 2 structures.

Iterative rounds of refinement formulate inference as a form of maximum likelihood.[36]

The standard of cryo-EM map reproductions is mostly reliant on regularization. These methods capitalise on earlier knowledge of domain to eradicate unwanted model complications and avoid overfitting. Regularization is necessary in cryo-EM refinement to lessen the effects of imaging and sample disturbance so that the ultimate 3D density contains only the protein signal.[37]

For current refinement algorithms, an explicit regulator in the form of a shift invariant linear filter (for more details on the meaning of shift invariance and linearity, see Mathematical Supplement), is used, which is typically obtained from a Fourier shell correlation (FSC).[38] These kinds of filters smooth the 3D structure using the same kernel and hence it leads to the same degree of smoothing in all directions. Since FSC usually reports the average resolution of the map, these kinds of filters typically under- and over-regularize specific regions, thereby bringing about dominance of noisy effects in some parts and a lack of resolvable details in others.[39] Since the regularization parameters fixed by the FSC in normal refinements are common to both the half maps, these are not fully independent. Initial refinements are performed without any symmetry. Subsequently, a job called 'symmetry expansion' available in Relion can be exploited to impose the symmetry followed by further improvements. This symmetry expansion job is not needed in cryoSPARC.

---

[36]https://sites.stat.washington.edu/jaw/COURSES/580s/581/HO/LeCam-1990.ISIRev.pdf.

[37]https://www.nature.com/articles/s41592-020-00990-8.

[38]Ali Punjani, John L. Rubinstein, David J. Fleet & Marcus A Brubaker cryoSPARC: algorithms for rapid unsupervised cryo-EM structure determination, 2017 Mar; 14(3):290–296. doi: 10.1038/nmeth.4169. Epub 2017 Feb 6. Nat methods. Grant T; cisTEM: user friendly software.

[39]Ali Punjani, Haowei Zhang & David J. Fleet Non-uniform refinement: adaptive regularization improves single-particle cryo-EM reconstruction, Published: 30 November 2020.

Non-uniform refinement (NU-refinement)

The phenomenon stated in the last section is more pronounced with comparatively small proteins and membrane proteins that are endowed with considerable non-uniform rigidity and disorder across the molecule. In NU refinement algorithm a set of particle images and a low resolution *ab initio* 3D map is taken as an initial data. This data is randomly separated into two halves, each of which is separately used to obtain a 3D half-map. The nature of such "gold standard refinement",[40] makes it feasible for the use of FSC in the evaluation of map quality and for contrasting with existing algorithms. The most crucial aspect in NU-refinement is cross-validation (CV) regularization applied to each iteration of 3D refinement. Following the assumptions of the 'gold standard', the regularisation parameters are independently evaluated for each half of the map. On average, uniform refinement in cryoSPARC[41] is twice as fast as NU-refinement. The latter is slowed down due to optimisation and use of the adaptive regulator. For rigid small proteins, the default parameters in cryoSPARC work well. However, for small, flexible proteins some of the parameters need to be modified. For example, dynamic masking, which controls how far the mask is expanded in the near as well as the far distance can be made more loose. If it is too tight, then the mask often leads to artefacts and sharp dips in the FSC curve. The dynamic masking start resolution can also be decreased from the default value of 6 Å to 12 Å.

Local refinement

Sub-volumes within the total volume have to be chosen for refinement. Once this is done a cover or mask has to be constructed around the sub-volumes, as mandatory for local refinement. Exploiting UCSF Chimera,[42] such masks are generated. This mask must engulf

[40]https://pubmed.ncbi.nlm.nih.gov/22842542/.
[41]Ali Punjani; Non-uniform refinement: adaptive regularization improves single-particle cryo-EM reconstruction.
[42]https://guide.cryosparc.com/.

a part of the protein bigger than 150 kDa in mass. Use of smaller masks is discouraged since important information might be missing and insufficient to align all particles to the masked volume. As a result, overfitting might happen which is often viewed as artefacts in the density (manifested as streaks, shells or high density "blips").

Mask generation in Chimera

The initial step in forming a mask is to open the volume in UCSF Chimera and binarizing the volume. Utilising the perfect viewing level, the mask is fixed to a threshold value using the "vop threshold" command, by putting to zero all values below the threshold. This is implemented by the following command using a level example of 0.20:

vop threshold #<volume-spec-0> minimum 0.20 set 0

This sets all voxels with value below 0.20 to 0 and returns the result in a new volume with model number <volume-spec-1>.

After binarizing the complete structure, the sub-volume can be deleted from the binarized structure by using the volume eraser. It is useful to note that, apart from volume eraser tool, there are some other mask generation methods available in Chimera. For instance, the Segger tool in UCSFChimera can be used to algorithmically segment an input volume and collect the other segments until the region of interest is present in one segment. On the other hand, if an atomic model already exists for the part of the structure that one wants to refine, it can be opened in the same Chimera session as the original map and the 'Fit in map' kit can be used to align it. Subsequently, a new volume from the atomic model can be formed by the molmap command. This serves as a mask for further refinement. It is important to observe that in either case, the "vop resample command" has to be used to resample the segment/volume of relevance on the grid of the original volume. On induction of masks in cryoSPARC, it is crucial to use the volume tools job that permits us to expand, fill holes and/or soften the masks. The dilation radius and threshold parameter should be fixed at a value (in voxels) which yields an enlarged mask that just goes beyond the original volume.[43] The standardised volume used to construct a mask is low pass filtered.[44]

Methods of junk particle removal

Various approaches exist to eliminate junk particles. Heterogeneous refinement is a popular tool used in this book. For those examples where individual sectors appear quite similar, this approach is rather effective since it enables simultaneous classification of particles and is able to identify the classes or sectors. A disordered set, naturally occurring in nature is the genesis of junk class (see Fig. 10).

Alignment free 2D classification + repeat steps like in a loop (see Fig. 11)

Alignment free 2D classification is another possibility of junk particle elimination. The particles abstracted from the finest refinement

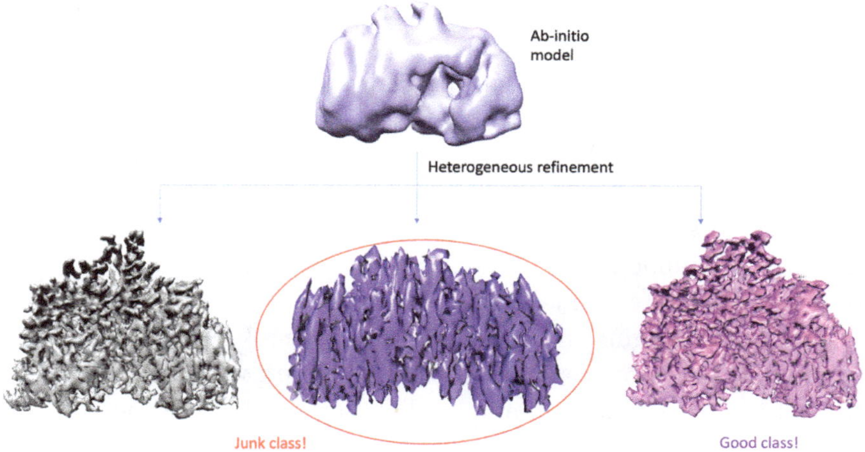

Figure 10: Heterogeneous refinement.

[43]https://guide.cryosparc.com/.

[44]Marina Serna, Frontiers, Hands on methods for high resolution cryo-electron microscopy structures of heterogeneous macromolecular complexes, PMID: 31157234 PMCID: PMC6529575 DOI: 10.3389/fmolb.2019.00033.

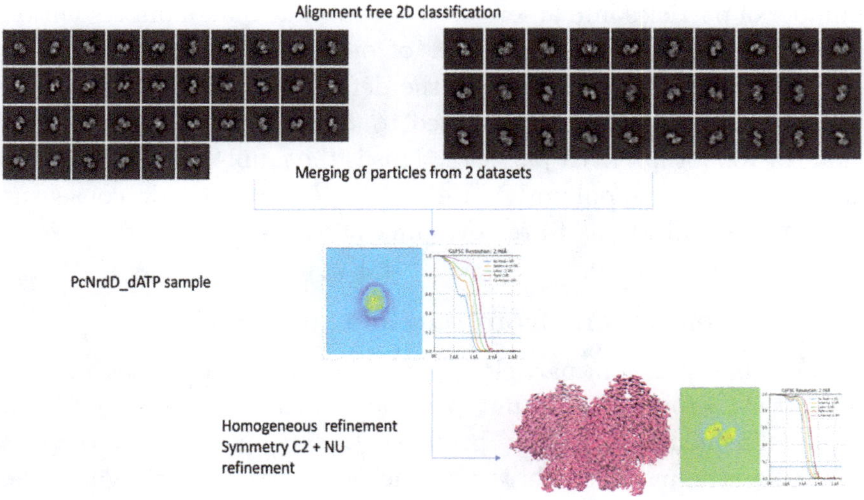

Figure 11:

job are put to a round of 2D classification. Alignment-free or reference-free 2D classification can be done at this step in Relion[45] A subsequent classification of particles is achieved by this step which helps in the removal of more junk particles. This iterative procedure is continued and improved classes are selected and processed further. The best particles obtained from this process are again employed for further refinement. After a few repetitions of these steps, it has to be verified whether the FSC resolution and the map quality get better or not. If there is an improvement from the previous resolution, the particles are subjected to a next round of 2D classification. This continues until there is no significant improvement in the $(n+1)$th iteration as found from the nth iteration.

For smaller particles, particles with low signal-to-noise ratio (SNR) or subsets with few distinct views, it is a good idea to run 2D classification with some modified parameters. Usually, during the first round of 2D classification, it is a good idea to classify the large

[45]Sjors Scheres, RELION; Implementation of a Bayesian approach to cryo-EM structure determination, *J. Struct. Biol. (JSB)*, PMID: 23000701 PMCID: PMC3690530 DOI: 10.1016/j.jsb.2012.09.006.

number of particles into at least 100 classes. The maximum resolution can be kept at 6 Å. The number of online EM iterations can be increased to about 40 instead of the default value of 20. The batch size per class can also be increased to 200 or 400 from the default value of 100 for low SNR particles. This is the number of particles per class to use during online EM iterations. We can also do consensus *ab initio* classifications to remove junk particles.

Focused refinement from particle subtraction

Undesirable signals in particle images are eliminated by the process of particle subtraction. Input particles which are employed for a local refinement job are the principal targets for this subtraction. Consequently, particle subtraction yields a particle stack where the unwanted signal (associated with the masked input volume) does not exist. This is effected when the job applies the mask to the volume, isolating the part of the density map containing the spurious signal. Next, the 3D alignment data of the particles is used and the projected signal from the masked volume is subtracted from each particle image.[46] Following the above statements, mask generation is done. Local refinement and particle subtraction are generally burdened with problems, especially related to mask padding and size. It is mandatory that the mask used in particle subtraction be softly padded, so that its edge slowly falls over space from 1 to 0. The desirable voxel value of the width of the soft padding is considerably greater than the expected resolution of complex. An effective thumb rule that is adopted provides the minimum mask softness as 5 × resolution (Å)/pixelsize(Å). After signal subtraction, the next natural step is local refinement of the sub-volume (whose signal remains in the subtracted particles).

Implementation of masks has been around for a long time during image processing to eliminate the solvent noise engulfing the molecular envelope. This is done especially during alignment

[46]https://guide.cryosparc.com/processing-data/all-job-types-in-cryosparc/local-refinement/job-local-refinement-beta.

and after 3D refinement where the gold standard FSC calculation is used.[47] To block artifacts in the Fourier domain during FSC estimations, masks must be endowed with soft, instead of sharp edges.

For refining a protein with given flexibility, local refinement is perfect. This process takes into proper consideration the motion between the masked sub-volume and the remaining structure. It is effective for refining sub-regions within the agreed volume, which becomes necessary when there is alleged movement between sub-units in the complex under consideration. In this technique, refinement of particle poses is done successively and recomputed by appropriately arranging them with the masked sub-volume that yields a better resolution in the final structure of the sub-volume. This method rests on picturing the sub-volume as a rigid body. <CryoSPARC.doc>

A crucial step in local refinement is fulcrum selection, which is discussed now.

The findings of local refinement depend on the fulcrum location, since it pinpoints the centre of the alignment search space over poses and translations. One possible location of the fulcrum for refining a sub-volume is the region within the boundary between the sub-volume and the remaining volume. Yet another possibility is to place fulcrum at the centroid of the mask which can be computed in UCSFChimera.[48] One example where I tried to use signal subtraction followed by local refinement is discussed below (see Fig. 12).

The search extents over rotations and translations also influence local refinement. The cherished choice may be found by performing multiple refinements with differing search extents. Beforehand knowledge about the extent of motion in the dataset can be used as an initial condition.

---

[47]https://www.researchgate.net/publication/230581939_Prevention_of_overfitting_in_cryo-EM_structure_determination and https://www.sciencedirect.com/science/article/pii/S1047847712002481?via%3Dihub.

[48]https://guide.cryosparc.com/processing-data/all-job-types-in-cryosparc/local-refinement/job-local-refinement-beta.

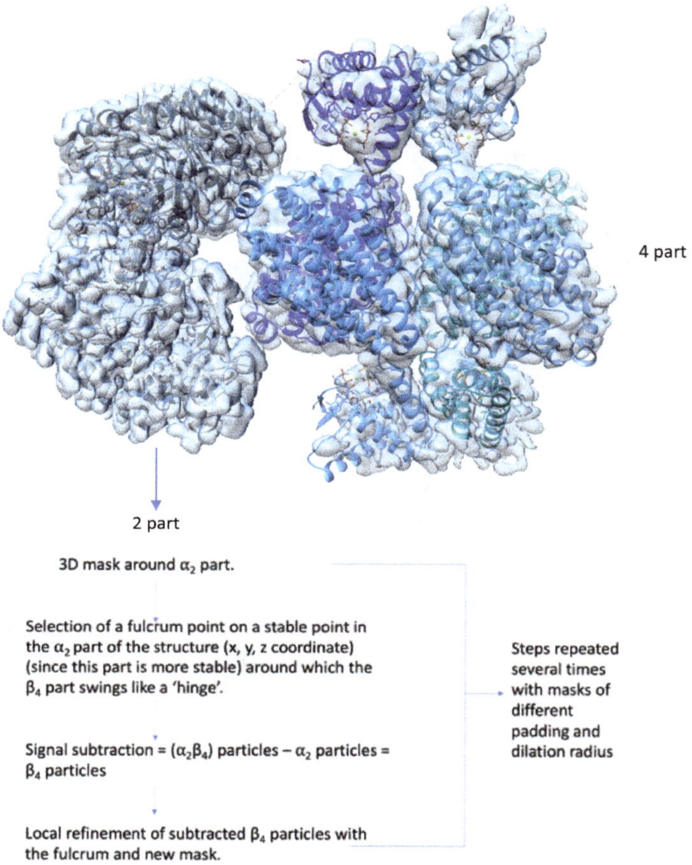

Figure 12: General workflow of local refinement.

Observing flexible molecular motions

Single particle cryo-EM is a fantastic tool for determining static structures of protein molecules, but the methods described so far are not great at modelling flexible proteins.To combat this problem, a new method called 3D variability analysis was introduced.[49]

[49]Ali Punjani and David J. Fleet; 3D variability analysis: Resolving continuous flexibility and discrete heterogeneity from single particle cryo-EM, *J. Struct. Biol.*, PMID: 33582281 DOI: 10.1016/j.jsb.2021.107702.

3D variability

3D variability analysis (3DVA) is a powerful tool in cryoSPARC used for exploring both discrete and continuous heterogeneity in cryo-EM datasets. In this kind of analysis, individual particle images are used to reconstruct a continuous family of 3D structures. As a whole, this family captures the multiple discrete and continuous conformation present in the data. The whole family can be thought of as the set of all possible 3D conformations that can be present in the sample. Multiple dimensions or attributes may be needed to characterise members of the family. For example, a subunit may associate/dissociate (1st attribute), while another subunit may bend (2nd attribute), while a ligand binds/unbinds (3rd attribute). 3DVA in cryoSPARC solves for the continuous family of 3D structures as well as assigning each particle to a position within the family (reaction coordinate).[50] 3DVA is a helpful way to visualize the flexibility and heterogeneity in a dataset.

3DVA is based on determination of the eigenvectors of the 3D covariance of a set of particle images. This method is also known as principal component analysis (PCA). A principal component is the same as an eigenvector of the covariance. Eigenvectors are literally linear directions in the space of 3D volumes in which there is significant variability in the dataset. They can also be thought as "trajectories" in the space of 3D structures along which the molecule exhibits conformational variability. Moving along a particular eigenvector gives all the different 3D conformations of a molecule. Multiple eigenvectors can be solved simultaneously, where these eigenvectors correspond to different types of variability. Each eigenvector itself is a 3D volume containing negative and positive values at each voxel.

In the case of discrete heterogeneity, we find multiple clusters in reaction coordinate space. Each cluster corresponds to a discrete conformation of the molecule, where there is greater probability of finding particles. Example: Sample reaction coordinate plot for LbNrdAB complex: is shown in Fig. 13.

---

[50]https://guide.cryosparc.com/processing-data/tutorials-and-case-studies/tutorial-3d-variability-analysis-part-one.

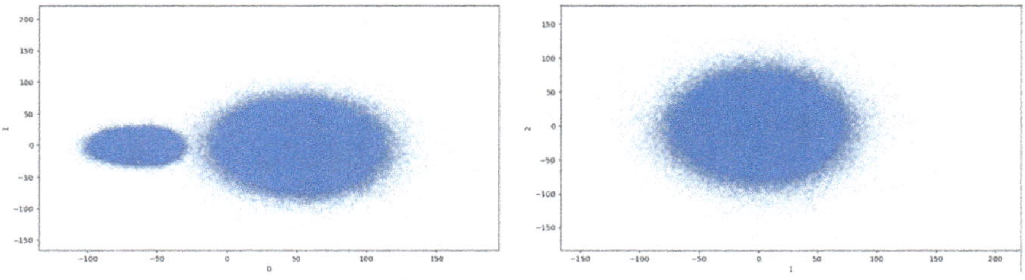

Figure 13: Sample reaction Coordinate plot.

The axes 0, 1 and 2 correspond to the three different components (eigenvectors) that were solved simultaneously using 3DVA. The left-hand plot shows coordinates for eigenvectors 0 vs 1, while the right-hand plot shows the same for eigenvectors 1 vs 2. The left plot is an example where we see discrete heterogeneity, as there are two clusters which implies two different conformations in the dataset. Eigenvectors 1 and 2, on the other hand, represent a more continuous or smoother flexing of the molecule.

In the case of non-discrete or continuous flexibility, we usually find a single cluster with significant flexibility in one or more dimensions, as seen in the right-hand plot. Some degree of flexibility is seen in both these plots. The components of variability will correspond to the flexibility itself and particles along the different directions of the reaction coordinate are in varying modes of flex. This flexibility can be visualized as a “movie” of volumes along the variability dimension.

Volume series

The particles and mask are taken from a consensus refinement job and used to run 3DVA.

Cluster mode

3D variability can also be used for separating discrete conformational changes. Classification or clustering is naturally difficult for cryo-EM datasets as there is more noise in this kind of data, more data points that need to be compared to one another etc. 3DVA takes account

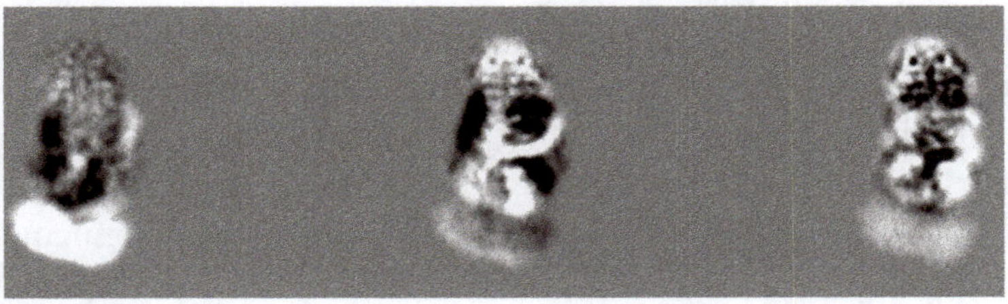

Figure 14: This shows a projection of each of the 3 modes of variability (eigenvectors) that were solved. Each one contains both black (negative) and white (positive) regions which imply the density changes.

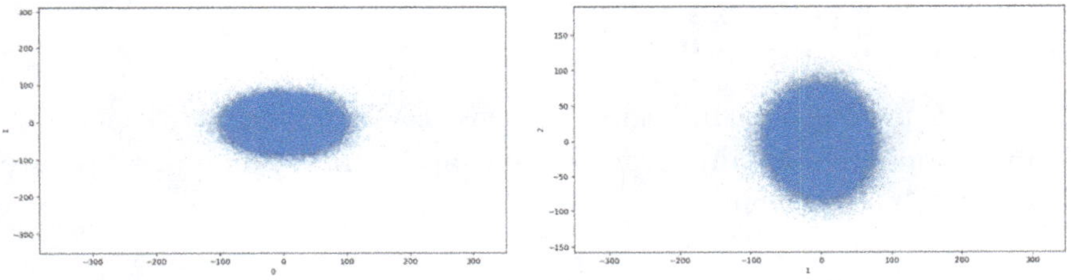

Figure 15: This figure shows the reaction coordinate distribution of particles as scatter plots between adjacent pairs of components (0 vs 1, 1 vs 2 etc.).

of these problems and makes clustering much simpler. It is based on the theoretical result: a linear manifold formed from eigenvectors of the data covariance (i.e. 3D variability components) will, under some mild conditions, span the subspace in which clusters lie. For cryo-EM heterogeneity, it means that there are discrete clusters present in a dataset and the first several 3D variability components will show us directly the difference between clusters, separating them as clearly as possible inspite of the heavy noise in the images. Thus, the problem of locating clusters becomes much simpler: they will show up as visual clusters, when we visualize the particles in the reaction coordinates (see Fig. 15).[51]

[51]Punjani; 3D variability analysis: Directly resolving continuous flexibility and discrete heterogeneity from single particle cryo-EM images, PMID: 33582281 DOI: 10.1016/j.jsb.2021.107702.

Multibody refinement in Relion

When biological macromolecules adopt many different conformations, traditional image processing approaches often lead to blurred reconstructions. If complexes are thought to be comprised of multiple, independently moving rigid bodies, multi-body refinement in RELION enables structure determination of highly flexible complexes, while at the same time characterising the motions of the complex. It can be applied on any cryo-EM dataset of flexible complexes which can be divided into two or more bodies, each with a minimum molecular weight of 100–150kDa.[52]

Post-processing tools

Different post-processing tools in cryo-EM software packages can bring about further improvement in map quality. Some of these are discussed below briefly:

Coma free alignment

Coma free alignment is a method which eliminates coma aberration and is important for high resolution electron microscopy. New algorithms for coma free alignment have been devised on the basis of beam tilt induced astigmatism.[53]

Global CTF refinement

Very high resolution cryo-EM structures need correction for electron-optical aberrations and microscope misalignments which result in "high order" terms in the CTF. These higher order terms (corresponding to beam tilt, trefoil, spherical aberration, tetrafoil) can be detected only at very high (atomic or near atomic scale) resolutions and cannot be estimated by straightforward measurements on the microscope. Hence, the strength of each of these aberrations must be estimated from single particle data itself, by refining the

[52] https://pubmed.ncbi.nlm.nih.gov/33368003/.
[53] K. Ishizuka, Ultramicroscopy V 55, Issue 4, Oct 1994, p 407–418.

corresponding CTF parameters using a high resolution reference map. This process of high order aberration estimation and correction was first introduced in RELION 3.1.[54]

Microscopes also occasionally show magnification anisotropy. This anisotropy results in the micrographs being slightly distorted by a linear transformation (or "stretch") in the image plane. Unlike the presence of higher order aberrations, anisotropic magnification cannot be corrected by better microscope alignment, but can be estimated by projection matching using a high quality reference map. High resolution signal is needed to estimate anisotropy and usually correcting this anisotropy will only improve maps that have already achieved high resolution. Images which are collected on a related microscope will have related CTF parameters for higher order aberrations and anisotropic magnifications. Images that are related (same grid, same image shift position etc.) are grouped into "exposure groups" so that they can all be refined at once with more signal.[55]

Local/per-particle motion correction

As stated earlier, during cryo-EM data collection, motion in the sample can occur due to reasons like stage drift (the sample as a whole moves) or there can be deformities in the sample as a result of the energy deposited by the electron beam (anisotropic deformation). It is fundamentally important to correct this motion. Local motion correction uses previous knowledge about particle locations within the micrographs to carry out motion correction on individual particles. Sometimes, this can lead to marginal improvement in the resolution estimation of final structure than what is seen from patch motion correction. As local motion correction needs the particle location, it is a good idea to go back and use local

[54]Estimation of high-order aberrations and anisotropic magnification from cryo-EM data sets in *RELION*-3.1, J Zivanov, T Nakane, SHW Scheres, 2019, https://doi.org/10.1107/S2052252520000081.

[55]https://guide.cryosparc.com/processing-data/tutorials-and-case-studies.

motion correction once the initial round of picking and extraction is complete.

Local CTF refinement (or per-particle defocus)

Post processing is defined as the modifications that we make to the map that are independent of any properties of the microscope or sample.

Per-particle refinement of CTF parameters and corrections of estimated beam tilt in RELION provide better resolution reconstructions when particles are at different heights in the ice, and/or coma-free alignment has not been optimal.[56] Local CTF refinement as implemented in cryoSPARC is quite a straightforward optimization process of finding the optimal per-particle defocus for each particle in a dataset. In cryoSPARC, local CTF refinement requires aligned particle images and a 3D reference (two half-maps), ideally already at high resolution. The experimental particle images are compared against the 3D reference from their half set, from the best known pose at various defocus levels and from here the best defocus is selected. The optimal defocus is usually the height of the particle in the sample/ice. Local CTF refinement works best for larger, highly rigid, high quality samples that have already reached high resolutions (better than 4 Å). Usually it is a good idea to try local CTF refinement on every dataset and then use a homogeneous (gold standard) refinement to estimate whether the overall resolution increased or decreased.[57]

Histograms showing the change in per-particle defocus across all the particles in the half-set indicate the total amount of deviation from the input defocus parameters that was achieved through CTF refinement. This histogram usually has a peak near zero and does not have heavy tails (see Fig. 16). Having heavy tails and the presence of many particles having optimal defocus values at the end of search range indicates that the defocus refinement was not very accurate.

---

[56] https://discuss.cryosparc.com/t/ctf-refinement-error-in-non-uniform-refinement/15318.

[57] https://guide.cryosparc.com/processing-data/tutorials-and-case-studies/tutorial-ctf-refinement.

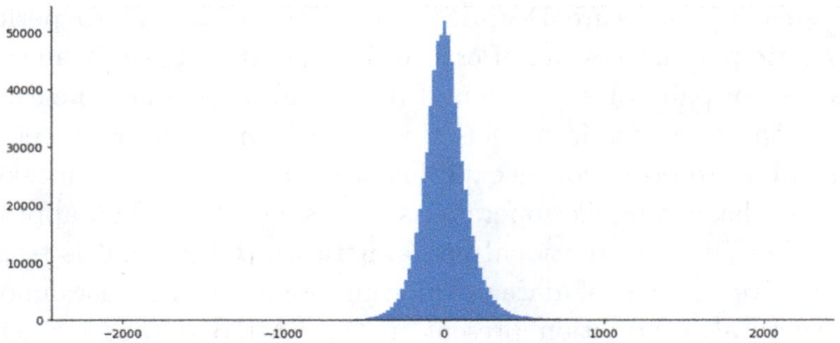

Figure 16: Defocus change across all particles split A (PcNrdD_dATP dataset).

Map improvement tools

LocScale algorithm

A method, known as LocScale, was developed which uses the radial structure factor from an atomic model that has been refined and used as a restraint for local sharpening of maps. It is a method for local map sharpening.

Cryo-EM map improvement using machine learning

Cryo-EM maps provide valuable information for protein structure modelling. However, due to the loss of contrast at high frequencies, these maps need to be post-processed to improve their interpretability. Approaches to compensate for this contrast loss and improve map visibility at high-resolution are important. This process is usually referred to as 'sharpening'. Previous sharpening approaches for cryo-EM maps were based on global B-factor correction. These methods included RELION postprocessing[58] or phenixautosharpen.[59] Most popular approaches based on global B-factor correction suffer from limitations in that they ignore the heterogeneity in the local map quality which reconstructions tend to exhibit. To overcome these

[58]https://academic.oup.com/bioinformatics/article/36/3/765/5554698.
[59]https://phenix-online.org/.

problems, the software DeepEMhancer was developed to perform automatic post processing of cryo-EM maps. It has been trained on a dataset of pairs of experimental maps and maps sharpened using their respective atomic models through which DeepEMhancer has learned how to post process experimental maps performing masking-like and sharpening-like operations in a single step. The approach used relies on a convolutional neural network (CNN) that is trained on massive chunks of data, exploiting some of the vast amount of structural information present in the EMDB database[60] which mimics the local sharpening effect of the locScale algorithm. Loc scale is a map sharpening program that exploits previous information from a refined atomic model to enhance contrast of cryo-EM maps. Most importantly, DeepEMhancer does not require any input atomic model to function. This was based on an end-to-end U-net architecture[61] trained in a supervised manner. Training was performed initially using pairs of input maps and target maps, consisting of experimental cryo-EM maps and tightly masked LocScale post processed maps. LocScale was chosen to produce targets because it makes use of atomic model information which tends to produce better quality results.[62] DeepEMhancer post processing carries out a non-linear transformation of the experimental volume that produces a set of effects which can be broadly classified as masking or denoising and sharpening-like feature enhancement (see Fig. 17).

Deep EM map resolution estimation by DeepRes

The program DeepRes is used to estimate the resolution of cryo-EM maps. It is based on deep learning 3D feature detection. DeepRes is completely automatic and parameter free and avoids limitations of most current methods, such as their insensitivity to enhancements owing to B-factor sharpening. This is an issue which has been neglected in the cryo-EM field for a long time. This way, DeepRes can be applied to any map, detecting subtle changes in local quality after the application of enhancement processes like isotropic filters

[60]https://www.biorxiv.org/content/10.1101/2023.10.03.560672v1.full.pdf.
[61]https://blog.aiensured.com/end-to-end-guide-for-u-net-model/.
[62]https://www.nature.com/articles/s42003-021-02399-1.

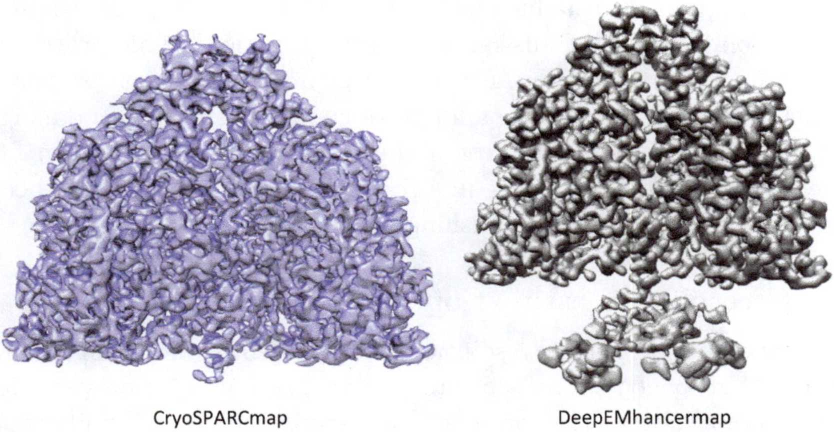

Figure 17: Map improvement with deepEMhancer software (PcNrdD_ATP dataset).

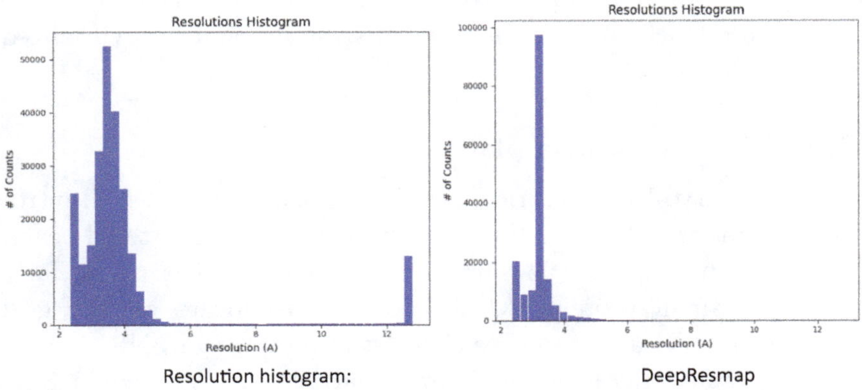

Figure 18: Resolution histogram.

or more complex procedures like model based local sharpening, non-model-based methods or denoising (see Fig. 18).

Bayesian polishing

This is a new method to estimate trajectories of particle motion and the amount of beam damage in cryo-EM. The motion present in the sample is modelled through a Gaussian process regression. A known set of data points, usually called the kernel, is used to predict a new set of points using a Gaussian (quadratic function). This process is called Gaussian regression. Finally, an algorithm has been proposed which estimates the average amount of cumulative radiation damage as a function of radiation dose and spatial frequency, and then fits the B-factors to that damage in a robust way. This method has been implemented as Bayesian polishing in RELION-3.[63]

Local resolution estimation

The introduction of local resolution measurements have changed how cryo-EM reconstructions are interpreted. Local resolution provides information about the presence of heterogeneity, flexibility and angular assignment errors and use it as a tool to help in modeling. Local resolution estimation is carried out both in cryoSPARC and Relion. In cryoSPARC, half maps of a refined volume are usually given as input. Output is given as map_locres which gives the resolution of each voxel of the original structure and it is visualized in UCSFChimera.[64]

Cryo-EM with phase plate

Sometimes cryo-EM is carried out with a phase plate, which introduces an additional phase shift of the un-scattered vs scattered beam. It is used to give more low frequency contrast and is particularly useful for small proteins which are hard to locate on the grid (<= 100 kDa). It also improves contrast in (even lower SNR) images of tomographic tilt series and fewer particles are needed in SPA and sub-tomogram averaging. However, the rule of thumb is to try without the phase plate first as both data collection and CTF correction are more complicated with phase plate.

[63]https://relion.readthedocs.io/en/release-3.1/SPA_tutorial/Polish.html.
[64]https://www.cgl.ucsf.edu/chimera/.

Chapter 4

Small Angle X-Ray Scattering

SAXS is an analytical technique which measures the intensities of the X-rays scattered by a sample as a function of the scattering angle. Measurements are usually made at very small angles which are typically in the range of 0.1 to 5 deg. From Bragg's law it can be seen that with decreasing scattering angle, increasingly larger features are probed. A SAXS signal is usually observed whenever a material contains structural features on the length range of nanometers, typically in the 1–100 nm range.[1]

Biological SAXS is a small angle scattering method for structure analysis of biological materials. SAXS and SANS (Small angle neutron scattering) are two complementary methods often known as small angle scattering (SAS). This method is capable of delivering structural information in the resolution range between 1 and 25 nm. For biological applications, SAS is used to determine the structure of a particle in terms of average particle size and shape. Usually the biological macromolecules are dispersed in a liquid and we can get information about the surface to volume ratio. This method is accurate, mostly non-destructive, and requires minimal sample preparation. However, biological macromolecules are sensitive to radiation damage.

In a SAXS experiment, a coherent, monochromatic X-ray beam hits the sample and the radiation which is scattered at low angles, typically a few degrees, which is recorded by a detector attached to

[1]https://www.malvernpanalytical.com/en/products/technology/Xray-analysis/Xray-scattering/small-angle-X-ray-scattering.

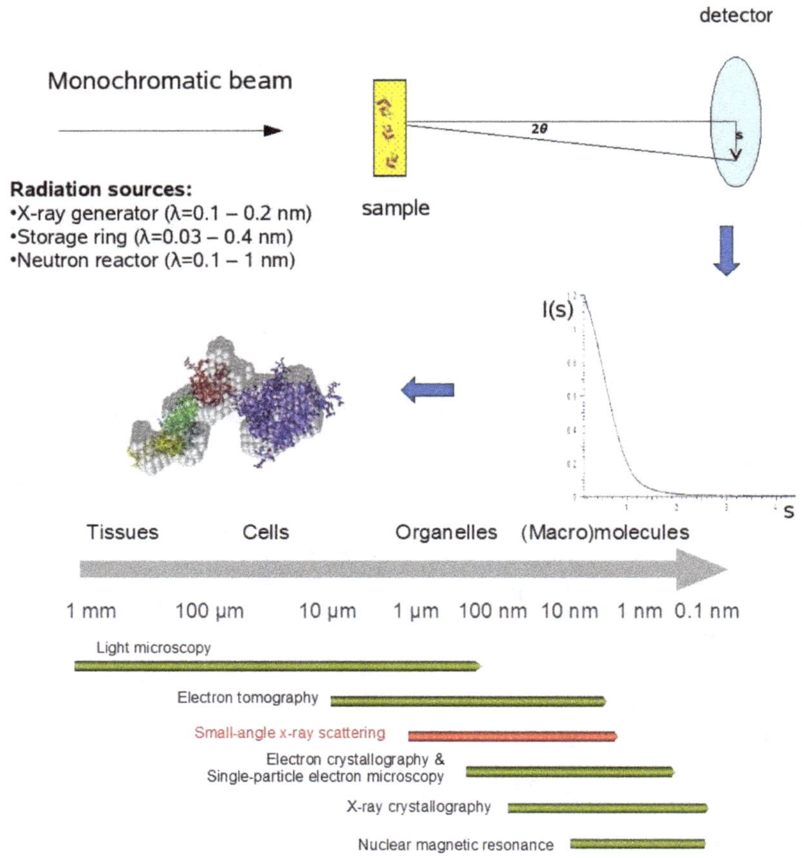

the computer. The scattering which originates from the electrons in the sample is mostly isotropic and radial averaging of 2D scattering pattern yields a 1D intensity curve.[2] The signal comes both from the macromolecules and the buffer and also the surrounding container. These are the unwanted contributions which are removed by background subtraction. Thus, the subtracted SAXS profile gives the intensity from the macromolecule as a function of the scattering angle. This scattered intensity is usually a faithful reflection of the structure of the macromolecule of interest for dilute solutions, if the contributions from aggregates etc. are negligible. Size exclusion

[2]M. Graewert, D. Svergun, Impact and progress in small and wide angle X-ray scattering (SAXS and WAXS), Current opinion in structural biology, 1 OCT 2013.

chromatography can be implemented in conjunction with SAXS (SEC-SAXS)[3] to separate the macromolecule or complex of interest from aggregates.

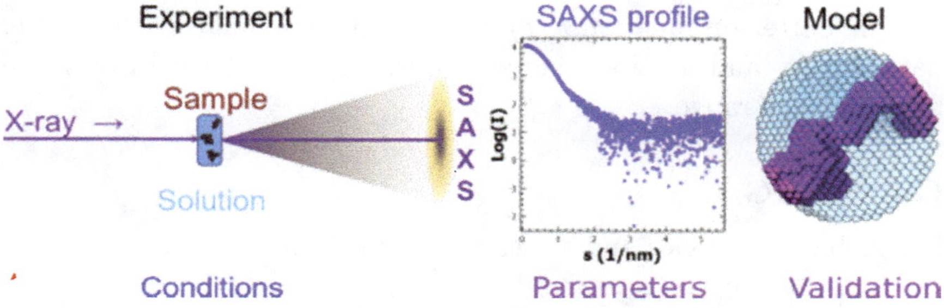

Methods, Developments and applications of small angle X-ray scattering to characterize biological macromolecules in solution.[4]

$P(r)$ function

In crystallography, we see that the Patterson function $P(r)$ corresponds to a map of distances between pairs of atoms in the structure. On the other hand, spherical averaging in the solution scattering experiment results in the loss of any directional information in the Patterson function, $P(r)$ which actually corresponds to the distribution of scattering centre pair distances.

$P(r)$ has a more intuitive relationship to shape than the scattering profile and can be expressed mathematically as a sum or integral:[5]

$$P(r) = \sum_{i,j} \Delta\rho(r_i)\Delta\rho(r_j), \quad \text{for all } r = |r_i - r_j|$$

or

$$P(r) = \iint_{r_{i,j}}^{r_{max}} \Delta\rho(\vec{r}_i)\Delta\rho(\vec{r}_j)d\vec{r}_i d\vec{r}_j$$

[3] David G and Perez, J (2009) J. Appl. Crys. Vol. 42, 892–900.

[4] Stefano de Vela, Dmitri Svergun, Current research in structural biology, Volume 2, 2020, Pages 164–170.

[5] For further details, see, Patterson, A.L. (1934), A Fourier series method for the determination of the components of interatomic distances in crystals; *Phys. Rev.* 46 (5) 372–376.

where r_{max} is the maximum atom pair distance. In the above equation the summation appears for a discrete distribution while the integral is valid for a continuous distribution. Their relationship is given by ($\sum \rightarrow \int dr$).

The Patterson function $P(r)$ is powerful for estimating the shape, data quality and for assessing self-consistency of the data over the measurement range.

Guinier RG

For globular particles and at relatively small angles,

$$I(q) = I(0)e^{\frac{-q^2 R_g^2}{3}}$$

where q is the scattering vector and Rg is the radius of gyration of the scattering particle. If the particle is spherical, the Guinier approximation holds for $qR_g < 1.3$. As particles become more asymmetric, the approximation limit narrows.

A plot of $\ln I(q)$ vs q^2 (in the guinier) region will give a straight line with an Intercept of $I(0)$ and a negative slope.

From Jill Trewhella's Lecture notes

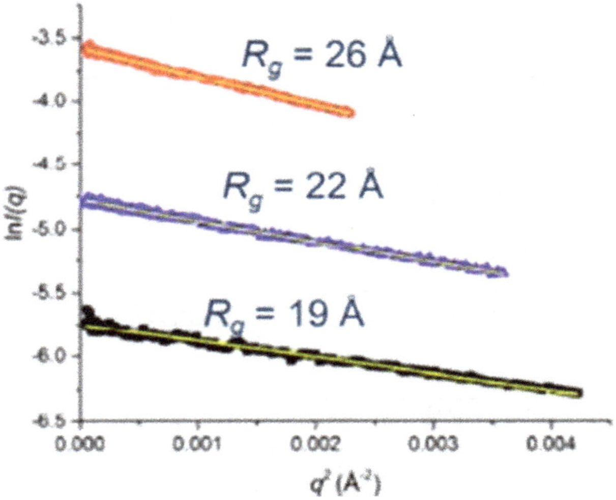

Porod Law

It is a description of the decay of the scattering profile at high q. For a folded particle with a sharp boundary (i.e. flat surface), Porod's law predicts the scattering intensity:

$$I(q) \sim Sq^{-4}$$

where S is the surface area of the particle. It is important to note that biomolecules do not generally have sharp boundaries and their state of "foldedness" varies.

Kratky plot

The Kratky plot $q^2(I)q$ vs q is used to assess the degree of compactness or "foldedness" of the scattering particle.

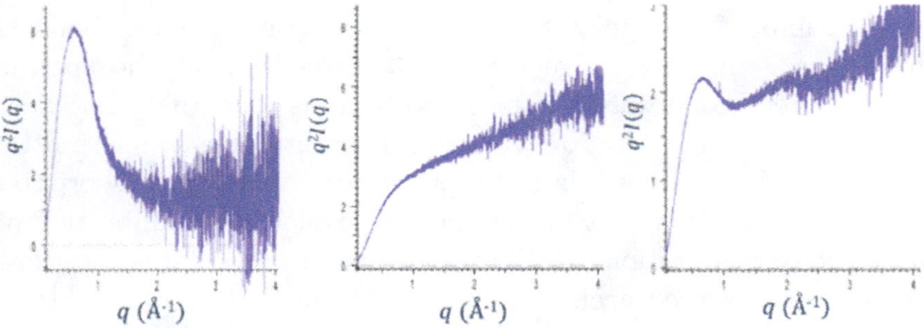

Folded Globular   Unfolded   Partially unfolded

The structural parameters which are inherent to the scattering profile are as follows:

$P(r)$, Rg, Particle mass (Mr) and volume, information on the particle surface.

SAXS sample preparation

Samples for solution scattering experiments can often be prepared at conditions which closely match the ones typically used for other experimental biophysical techniques for crystallography or NMR. The sample should contain the biomolecule of interest and a buffer is required which identically matches the composition of the solvent within the sample. The contents of the buffer can vary but not without restrictions. To bring about minimization of X-ray absorption and subsequent production of free radicals (mainly OH), the buffers should be free from high Z elements as far as possible. Also, the concentration of phosphates should be kept to a minimum (below 20 mM). Presence of detergents has to be controlled. Buffer compositions should be optimized as far as possible to ensure monodispersity of the samples at the scattering measurement conditions. Frequently, DTT at 5–10 mM or TCEP at 1–2 mM are used as additives for SAXS buffers in the cases where the macromolecules are compatible with them. When DTT and TCEP can no longer be used, radiation damage organic buffering agents like Tris or HEPES or free radical scavengers like glycerol, ascorbate, ethylene glycol, or sucrose can be used. In the cases where samples are stable at high ionic strengths, salts can be used to decrease the long-range electrostatic repulsion at the specific measurement conditions. In most cases, it is seen that the exact wavelength used in the X-ray scattering measurement makes little impact on the measured data except for decreased X-ray absorption and radiation damage at higher energies. Following equilibration of buffer the concentrations of all stock samples used for data collections should be measured accurately, preferably by UV-VIS absorption. The sample concentration needs to be determined with an error not exceeding 10% for accurate determination of the molecular weight of the macromolecule via SAXS.

For the stock solutions, concentrations not exceeding 5–10 mg/ml should be used.

It is also highly recommended to carry out sample preparation close to the measurement time, to minimize sample aggregation and to ensure best sample/buffer match. For preparing the macromolecular stock solution of the sample, the likely concentration range of scattering measurements has to be kept in mind. The preparation here also has to take into account the amount of material that would be used during the whole measurement process, considering the volumes to be loaded during data collection.

The last steps in the sample preparation process should include quality assurance procedures which help to establish sample purity and the absence of high molecular weight aggregates. Acceptable methods for assessing sample quality are SDS PAGE, native gel filtratioan and dynamic light scattering (DLS). Along with the samples of interest it is also recommended to prepare several standard samples. Standards can be useful in two ways: the standards allow us to test for proper operation of the instrument and to correct any problems if found before the samples are loaded. Secondly, well behaved standards which have established oligomerization states and precisely measured concentrations can be used to determine the molecular weights and aggregation states for the samples of interest. The most commonly used standard is hen egg white lysozyme at pH 4.0–4.4.

SAXS data collection

Solution X-ray scattering data can be acquired using either lab-based instruments or synchrotron beam lines. Lab-based instruments usually operate at fixed specific wavelengths, typically Cu-K alpha at 8 KeV, while synchrotrons allow variation of the incident X-ray energy, usually between 7 and 20 keV. Lab-based sources allow the possibility of taking immediate measurements when the sample and instruments are available at the expense of relatively long data collection times (between 30 min and several hours). Since lab based instruments usually have moderate flux, radiation damage is usually less of a problem here compared to synchrotron instruments.

Synchrotron beam lines give the possibility to vary the energy of incident X-ray photons, which is often useful to adjust the q range of the acquired data and also to radiation damage with higher incident X-ray energies. For both lab based and synchrotron data collections, the space between the sample and detector has to be kept in low vacuum. The data collection that has been used here have been done at the synchrotron ESRF, Grenoble. Setup and planning of the synchrotron SAXS data collection starts from selection of the sample/detector distance(s) and the incident X-ray energy in order to cover the entire q range. The main advantage of using synchrotrons is the extremely high photon flux that can be achieved with modern undulator beam lines.

Scattering data collections should include measurement of the scattering from the cell filled with buffer and the sample, with the empty cell and cell filled with water. Both the sample and buffer data collections should be done with the same cell, identically positioned and cleaned thoroughly between measurements to remove any macromolecular deposits from cell walls. It is usually recommended to start with buffer measurements and follow it with the sample using the same exposure times. When calculating a concentration series on the same sample, data subtractions are most accurate if separate buffers are used before each sample dilution. SAXS measurements should include a series of concentrations for evaluating oligomerization and interparticle effects. At least 3 to 5 concentration points should be acquired with dilutions by a factor 2. For proteins with molecular weight above 100 kDa, highest measured concentration of 1–2 mg/ml is recommended while for smaller macromolecules, the highest concentration can be in the range of 5–10 mg/ml.[6]

SAXS data processing using various programs

PRIMUS

In the process of SAXS data processing, the program PRIMUS performs the necessary manipulations with experimental small angle scattering data files like: averaging, subtraction, merging, extrapolation to zero concentration and curve fitting. It thus evaluates the integral parameters from Guinier and Porod plots such as radius of gyration, Porod's volume, zero intensity and molecular weight. This program allows the user to make interactive manipulations with data that is convenient for users and saves a lot of time.

[6]Sample preparation, data collection and preliminary data analysis in biomolecular solution X-ray scattering, Alexander Grishaev, *Curr. Protoc. Protein Sci.* 2012, Nov, Chapter: 17.14.

AutoRg

AutoRg works with experimental data files in standard ASCII format. First, the program usually selects the data range suitable for Guinier approximation. For this, the initial portion of the data is analysed and the ranges showing unreasonable upward and downward trends (caused by beam stop or strong background near primary beam) are discarded. After this the data range, where the scattering intensity decays by an order of magnitude is taken.

Gnom

Program Gnom is used to calculate the distance distribution function $P(r)$. The *rmin* and *rmax* values define the bounds for the possible particle sizes present in the sample. By default, these values are equal to zero. There is also an option used to fix zero values of the $P(r)$ function at this point.

DAMMIF

DAMMIF is a method for rapid *ab-initio* model generation or shape determination in small angle scattering. It is a re-implementation of the well known bead modeling program DAMMIN. In the process of bead modeling, a particle is usually represented as a large number of densely packed beads inside a search volume. Each bead usually belongs to either the particle or the solvent. We start here from

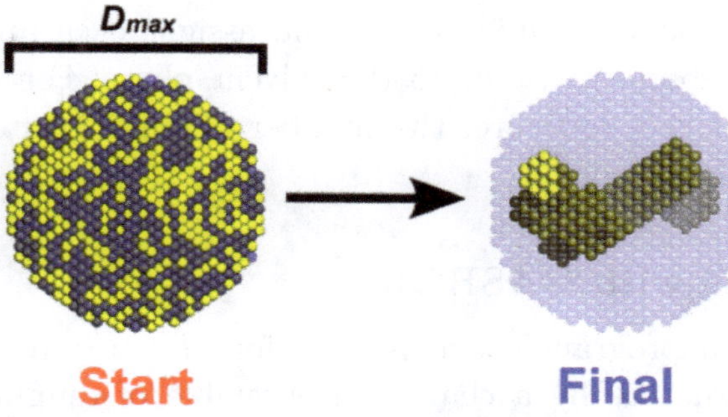

http://airen.bcm.umontreal.ca/biostruct/SAXS_tutorials/modelling_with_dammif/

an arbitrary initial model and use DAMMIF which uses simulated annealing to construct a compact inter-connected model yielding finally a scattering pattern that fits the experimental data.

Damaver

Damaver is a program to align *ab initio* low resolution models e.g. those provided by DAMMIF, DAMMIN and GASBOR then select the most probable model and provide averaged, filtered model which is suitable for search volume input for DAMMIN (also known DAMSTART model). If many models are supplied in PDB format, the program superimposes all possible pairs by applying one of the possible methods of superposition. The most representative model from here is determined and outliers are excluded from further calculations. The aligned models are taken and remapped onto a grid of densely packed beads to compute a frequency map.

DAMMIN

The program DAMMIN uses a method to restore *ab-initio* low resolution shape of randomly oriented particles in solution (e.g., biological macromolecules) from small angle X-ray scattering. A search volume is taken having a sphere of relatively large radius R which encloses the particles and is filled with N-densely packed spheres of radius r which is referred to as dummy atom. As the spatial positions are fixed, the dummy atom model is completely described by a vector X with N components which in the end assigns each dummy atom either to the solute phase or to the solvent phase. For an effective description of the structure, the number of dummy atoms should reach a few thousands.[7]

Bead model with GASBOR

GASBOR is a program which is used for *ab initio* reconstruction of protein structure by a chain like ensemble of dummy residues.

[7]Dammin Manual, D.I Svergun (1999), *Biophysics J*, 2879–2886.

Usage of GASBOR is similar to that of DAMMIN or DAMMIF where most of the parameters have the same meaning. The biggest difference here is that the protein structure is not represented by dummy spheres (called dummy atoms in DAMMIN/DAMMIF) but it is represented by an ensemble of dummy residues (which correspond to average residue densities) placed anywhere in continuous space with a preferred number of close distance neighbors for each atom.[8]

Electron density map generation with DENSS

DENSS (DENsity from Solution Scattering) is an algorithm which is used for calculating 3D particle electron densities from 1D solution scattering data.

Detailed molecular structure determination is usually carried out by X-ray crystallography and in recent years with single particle cryo electron microscopy. One of the alternatives to crystallography which is very popular for studying dynamics is the usage of solution scattering. This technique involves scattering the X-rays off molecules floating in solution rather than arranged in crystal. This enables the molecules to move dynamically in their natural states which enables us to observe large scale conformational dynamics important in biological function. Algorithms for extracting 3D information from 1D experimental data have been developed which have enabled the reconstruction of low resolution molecular envelopes, outlines of the particle shapes.

DENSS on the other hand is a new algorithm which enables reconstruction of 3D electron density function of a molecule from 1D solution scattering data. For the first time, researchers can look inside these molecules which are floating in solution to understand the internal density variations rather than seeing only the envelope of the particle shapes. This added information can enable researchers to better understand molecular structure in solution, particularly visualizing large scale conformational dynamics. DENSS usually

---

[8]Determination of domain structure of proteins from X-ray solution scattering, D.I. Svergun, M.V. Petoukhov, and M.H. Koch, *Biophys J.* 2001 Jun; 80(6): 2946–2953. doi: 10.1016/S0006-3495(01)76260-1.

achieves this by expanding upon the mathematical technique in the field of imaging known as iterative phase retrieval. This method of iterative phase retrieval has been used in diverse fields like coherent diffractive imaging, astronomy, tomography etc. and it brings about the reconstruction of images from only 3D intensity information (i.e. amplitudes) while missing other 3D important information — the phase. Grant's method which is known as "iterative structure factor retrieval" expanded this by reconstructing not only the 3D phases but also the 3D amplitudes by only taking into account the 1D averaged intensities from the solution scattering experiment.[9]

[9]https://www.nature.com/articles/nmeth.4581.

Chapter 5

Ribonucleotide Reductases (RNRs)

Ipsita's work in a nutshell on RNRs

Ribonucleotide reductases (RNRs) are essential enzymes that catalyzes the synthesis of DNA building blocks in virtually all living cells. RNR inhibition precludes DNA transcription and repair, resulting in inability of cells to proliferate. RNRs are well-recognized targets for cancer treatment and antiviral agents. Since bacteria possess RNRs which are different from those of humans, it makes these enzymes excellent targets for the development of antibiotics. Our research concentrates on various bacterial RNRs. We aim to understand their mechanism of action in order to expand our knowledge in bacterial physiology and develop novel antibacterial treatments. We are particularly interested in the ATP-cones — evolutionary mobile protein domains regulating the activity of RNRs.

Being a structural biologist, Ipsita employed the most advanced methods to obtain high-resolution structures of bacterial RNRs and their ATP-cones. Ipsita mastered cryo-electron microscopy (cryo-EM), X-ray crystallography and small angle X-ray scattering (SAXS). Ipsita successfully employed all three techniques and their combination to solve the structures of RNRs from different bacteria, such as a human commensal pathogen Facklamiaignava and the marine bacterium Leeuwenhoekiellablandensis. Ipsita used cryo-EM to solve the structure of an anaerobic (oxygen sensitive) RNR from the bacterium Prevotellacopri, which causes inflammation in humans. Very little is known about anaerobic RNRs and how they are regulated. Since these RNRs are abundant in bacteria causing severe diseases, Ipsita's work is of great importance for being able to combat these bacteria. Ipsita was also involved in attempts to solve the structure of an RNR regulatory protein NrdR from *Escherichia coli*, in complex with DNA. The regulator NrdR is present in the majority of bacteria including severe pathogens such as Mycobacterium tuberculosis and Staphylococcus aureus. Obtaining high resolution protein structure enables scientists to understand the protein function in detail and potentially find reagents that will specifically affect them in a desired way. These are just a few examples of Ipsita's work, which she did in collaboration with scientists from Stockholm University, and there is of course much more to her research. Ipsita's work has already been published in important scientific journals, but much is still in the process of publication. Since processes in science are slow, research takes many years and never stands still, the importance of scientific findings can be fully appreciated many years after their discovery. Ipsita's work makes an important contribution to the field of bacterial RNR research. It will serve ground for upcoming discoveries and will help saving human Lives in the future.

Inna Rozman Grinberg, PhD,
Researcher,
Department of Biochemistry and Biophysics,
Stockholm University.

RNRs are key enzymes that mediate the synthesis of de oxyribonucleotides, which are the DNA precursors for DNA synthesis in every living cell. This enzyme converts ribonucleotides to deoxyribonucleotides, the building blocks for DNA replication and repair.

RNRs have contributed to the appearance of genetic material that exist today, being essential for the evolution of all organisms on earth.

The strict control of RNR activity and dNTP pool sizes is important as pool imbalances lead to higher mutation rates, genome instability and replication defects.[1]

All organisms require active DNA synthesis, prior to cell division. There should be a balanced supply of the different deoxyribonucleotide triphosphates (dNTPs).

[1]https://pubmed.ncbi.nlm.nih.gov/24809024/.

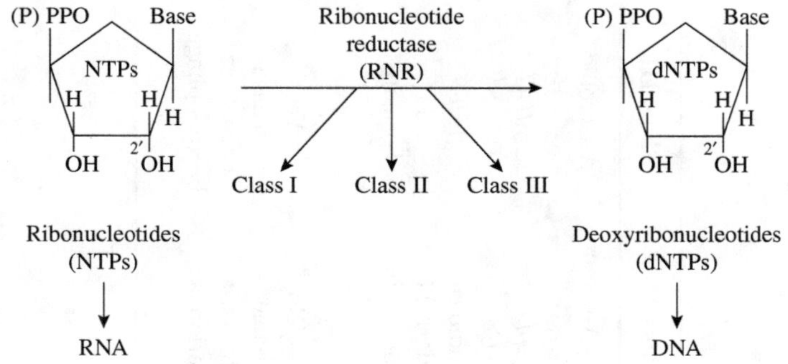

The only biochemical pathway for de novo dNTP synthesis is the reaction catalyzed through the enzyme ribonucleotides triphosphates (NTPs) into their corresponding dNTPs through reduction of the C2′ – OH.

Structure and mechanism of RNR is quite elaborate. It uses radical chemistry to catalyze the reduction of each NTP.

Three different RNR classes have been described — class I, class II and class III.

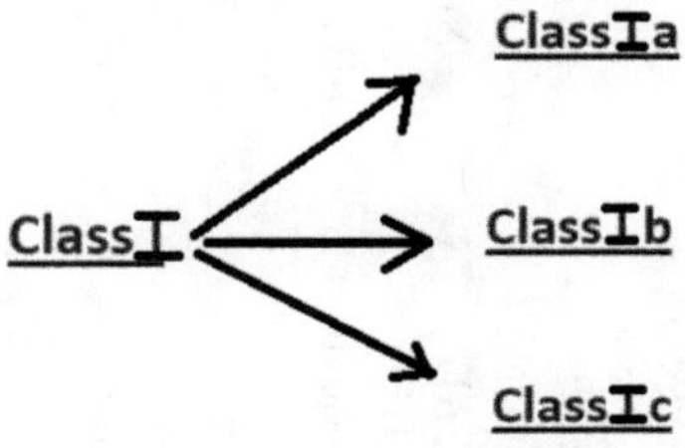

All 3 RNR classes → share a common 3D protein structure at the catalytic subunit and a highly conserved $\alpha\,|\,\beta$ barrel structure in the active site of the enzyme.

Also, the two potential allosteric centres (specificity and activity) are highly conserved among the different RNR classes, although in class Ib and some class II RNRs activity, alosteric site is absent.

Reduction of 4 different nucleotides (ATP/CTP/GTP/TTP) occurs at a single active site in each polypeptide chain, hence the tight regulation of dNTP levels is important for each dividing cell.

Unbalanced dNTP levels → could lead to increased mutation rates.

Properties	Class Ia	Class Ib	Class Ic	Class II	Class III
Oxygen dependence	Aerobic	Aerobic	Aerobic	Aerobic/Anaerobic	Anaerobic
Structure	$\alpha_2\beta_2/\alpha_6\beta_6$	$\alpha_2\beta_2$	$\alpha_2\beta_2$	$\alpha(\alpha_2)$	$\alpha_2 + \beta_2$
Gene	*nrdAB*	*nrdHIEF*	*nrdAB*	*nrdJ*	*nrdDG*
Metallocofactor "*In vivo*"	Fe^{III}-O-Fe^{III}	Mn^{III}-O-Mn^{III}	Mn^{IV}-O-Fe^{III}	*Co*	Fe^{II}-S^{II}
Cofactor assembly	YFAE	NrdI	Unknown	Unknown	IscF
Substrate	NDP	NDP	NDP	NDP/NTP	NTP
Reductant	thioredoxin, glutaredoxin	NardH-redoxin, glutaredoxin	Unknown	Thioredoxin	Formate
No. of allosteric sites	2	1	2	1/2	2
ATP inhibition	Yes	No	Yes	Yes/No	Yes
Distribution	Eukaryotes eubacteria Archaea bacteriophage virus	Eubacteria	Eubacteria	Archaea eubacteria bacteriophage	Archaea eubacteria bacteriophage

RNR activity is controlled at 2 different levels:

1. **Substrate specificity** — in which the binding of different nucleotides results in the reduction of each specific NTP at the active site and
2. **Enzyme activity** — in which the binding of ATP or dATP respectively activates or inhibits enzymatic activity.

Here, we focus on bacterial RNR: → class I RNR:-

a) Best known and most studied enzyme.
b) They comprise 2 homodimeric subunits, namely R1 (or α) and R2 (or β). The α subunit has the catalytic subunit containing the active site where nucleotide reduction occurs and 2 allosteric sites involved in the allosteric regulation of substrate specificity and general activity.

Active form of RNR in eukaryotes and prokaryotes comprises 2 proteins (R1+R2 or $\alpha+\beta$) associated in a dimeric or other oligomeric forms, such as $\alpha_n\beta_m$ (where n can be 2, 4 or 6 and m can be 1, 2, 3 or more).

→ Class I RNR → can be further subdivided into Ia, Ib and Ic, based on the type of metal centre required to generate the protein radical.

→ The nrdAB genes encode classIa enzymes, which require a di-iron centre (Fe^{III}-O-Fe^{III}) in the NrdB(β) subunit to generate the tyrosyl radical. → The nrdHIEF genes encode the class Ib operon, with NrdE and NrdF encoding the α and β subunits, respectively, NrdI encoding a Flavodoxin and NrdH encoding a glutaredoxin like protein. NrdI encodes a specific protein involved in the biosynthesis and maintainence of the active metal centre and NrdH is the sp. e^- donor for the NrdEF enzyme system.

→ Class Ib RNR → Has a dimanganese centre (Mn^{III}-O-Mn^{III}) to generate Tyrosyl radical *in vivo* can also be generated with a di-ferric centre.

Catalytic subunit NrdE → Also differs from that of class I RNRs because this enzyme lacks the activity site at the N-terminal region of the protein.

→ In all 3 class I RNRs → During catalysis the radical is formed in the β subunit and subsequently transferred to the large subunit (α) via a long-range radical transfer pathway, generating a thiol radical at the active site of the enzyme, where two cysteins are ultimately resposible for NTP reduction. Furthermore, all class I RNRs require the presence of D_2 for radical generation under aerobic conditions.

→ Class II RNR:-

Class II RNRs comprise a single α-chain polypeptide encoded by a single nrdJ gene.

→ This RNR class uses S-adenosycobalamine (AdoCob) to generate the cysteinyl radical that substitute the class I small protein (β subunit). This enzyme reaction does not require O_2 as this RNR class is completely O_2 independant.

Class II RNR → Have an allosteric specificity site but lack the allosteric activity site, similar to class Ib RNRs.

Exception!!

An exceptional class II RNR was identified in Pseudomonas aeruginosa (P. aeruginosa) and their enzyme differs from all known class II RNRs as this enzyme is split and encoded by 2 consecutive open reading frames, namely nrdJa and nrdjb seperated by 16 bp.

→ Class III RNR:-

Class III RNRs consist 2 homodimeric proteins encoded by nrdD and nrdG genes. NrdD is the large enzymatic catalytic subunit, harboring the active site and the 2 allosteric regulation sites, which respectively determine substrate specificity and activity.

Also the NrdG protein (known as activase) is responsible for generating the radical.

Class III RNRs require the binding of S-adenosylmethionine (SAM) to a 4 Fe–$4S$ metal center located in the NrdG protein for radical formation.

This interaction generates an extremely O_2 sensitive glycyl radical at the c-terminal domain of the NrdD protein.

Due to the high sensitivity of the glycyl radical and the potential metal centre oxidation in the presence of O_2, this RNR class is active only under anaerobic conditions.

Experiments with lactococcus lactis class III RNR have shown that NrdD alone catalyses the reduction of NTPs. After NrdD activation NrdG was no longer required and dissociated from the complex. Activated NrdD protein can catalyze several rounds of NTP reduction in the presence of formate as an e^- donor, with ATP as an allosteric effector and requiring Mg^{2+} and K^+ to stimulate the reaction.

Chapter 6

Solution Structure of the dATP-Inactivated Class I Ribonucleotide Reductase From *Leeuwenhoekiella blandensis* by SAXS and Cryo-Electron Microscopy*

Mahmudul Hasan[1], Ipsita Banerjee[1], Inna Rozman Grinberg[2], Britt-Marie Sjöberg[2] and Derek T. Logan[1]

[1] *Biochemistry and Structural Biology, Dept. of Chemistry, Lund University, Lund, Sweden*
[2] *Dept. of Biochemistry and Biophysics, Stockholm University, Stockholm, Sweden*

The essential enzyme ribonucleotide reductase (RNR) is highly regulated both at the level of overall activity and substrate specificity. Studies of class I, aerobic RNRs have shown that overall activity is downregulated by the binding of dATP to a small domain known as the ATP-cone often found at the N-terminus of RNR subunits, causing oligomerization that prevents formation of a necessary $\alpha_2\beta_2$ complex between the catalytic (α_2) and radical generating (β_2) subunits. In some relatively rare organisms with RNRs of the subclass NrdAi, the ATP-cone is found at the N-terminus of the β subunit rather than more commonly the α subunit. Binding of dATP to the ATP-cone in β results in formation of an unusual β_4 tetramer. However, the structural basis for how the formation of the active complex is hindered by such oligomerization has not been studied. Here we analyse the low-resolution three-dimensional structures of the separate subunits of an RNR from subclass NrdAi, as well as the $\alpha_4\beta_4$ octamer that forms in the presence of dATP. The results reveal a

*The chapter was originally published in Frontiers in Molecular Biosciences, Hasan M, Banerjee I, Rozman Grinberg I, Sjöberg B-M and Logan DT (2021) Solution Structure of the dATP-Inactivated Class I Ribonucleotide Reductase From Leeuwenhoekiella blandensis by SAXS and Cryo-Electron Microscopy. *Front. Mol. Biosci.* 8:713608. doi:10.3389/fmolb.2021.713608.

type of oligomer not previously seen for any class of RNR and suggest a mechanism for how binding of dATP to the ATP-cone switches off catalysis by sterically preventing formation of the asymmetrical $\alpha_2\beta_2$ complex.

Keywords: Ribonucleotide reductase, allosteric regulation, oligomerization, nucleotide binding, small-angle X-ray scattering, single particle cryo-EM

Introduction

The enzyme ribonucleotide reductase (RNR) catalyzes the reduction of ribonucleotides to deoxyribonucleotides. Being the only source of dNTPs for DNA synthesis and repair, RNR is essential for all but a very few living organisms. Since its discovery in the 1950s, RNR has continuously delivered surprises in terms of radical chemistry, allosteric regulation and molecular organisation, which have all been extensively reviewed previously (Hofer *et al.*, 2012; Ahmad and Dealwis, 2013; Torrents, 2014; Lundin *et al.*, 2015; Mathews, 2018; Greene *et al.*, 2020; Högbom *et al.*, 2020). The very large family of RNR enzymes can be divided into three major functional classes based on their radical generation mechanisms (Högbom *et al.*, 2020). All RNRs have a catalytic subunit with a 10-stranded α-β barrel fold, in class I called NrdA or α. Class I RNRs are activated by a radical generating subunit from the ferritin superfamily of all-α-helical proteins (NrdB or β). Class II RNRs, which can function both aerobically and anaerobically, have only a catalytic subunit and generate a 5′-deoxyadenosyl radical directly in the active site by homolytic cleavage of the C-Co bond in adenosylcobalamin. Class III, anaerobic RNRs also generate a 5′-deoxyadenosyl radical but do so by homolytic cleavage of S-adenosylmethionine by an accessory activase. This radical in turn generates a glycyl radical in the active site of the catalytic subunit. All three classes are unified by the transfer of these radicals to a cysteine next to the substrate in the active site.

RNR proteins are regulated at two levels: overall activity and substrate specificity. RNRs are active in the presence of ATP and downregulated by dATP through binding of these nucleotides to a small, approximately 100-residue domain known as the ATP-cone (Aravind *et al.*, 2000), most often found at the N-terminus of the catalytic α subunit. Over the past decade it has become evident

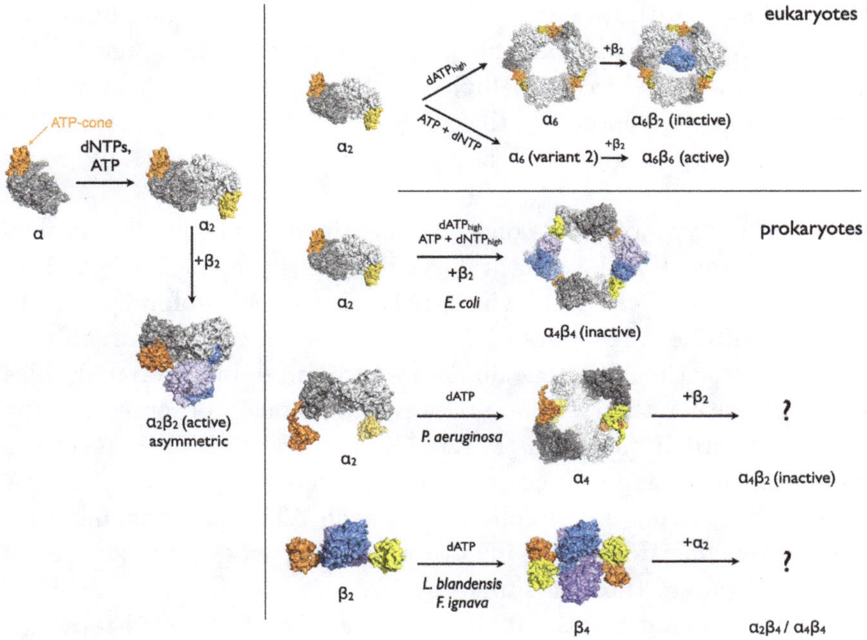

Figure 1: The variety of oligomers generated by binding of dATP to the ATP-cone in RNRs from different organisms. Low concentrations of dATP or other specificity-regulating dNTPs shift the monomer-dimer equilibrium towards dimers. Higher dATP concentrations induce the formation of inactive complexes of differing stoichiometries. The core domains of each NrdA subunit are shown in two shades of grey. The two ATP-cone domains in each dimer are shown in orange and yellow. The NrdB proteins are coloured in two shades of blue.

that the ATP-cone, in the presence of dATP, directs the formation of a variety of catalytically incompetent oligomers (Figure 1), all of which have in common that they break the sensitive chain of residues necessary for the proton-coupled electron transfer (PCET) that must occur upon each cycle of catalysis. In *E. coli*, two dimers of NrdA and two dimers of NrdB form a ring-shaped $\alpha_4\beta_4$ octamer in which the β_2 dimers are sandwiched between the α_2 dimers in such a manner that the PCET chain is broken (Ando *et al.*, 2011). In eukaryotic RNR, three dimers of NrdA form a ring-shaped α_6 hexamer (Fairman *et al.*, 2011; Brignole *et al.*, 2018) to which one β_2 dimer can attach (Rofougaran *et al.*, 2006; Fairman *et al.*, 2011;

Ando *et al.*, 2016), though the interactions with NrdB and how they break the PCET chain are not fully structurally characterized for this system. More recently, higher order oligomers such as helices have also been proposed both for class Ib RNRs with partial ATP cones (Thomas *et al.*, 2019) and class Id RNRs lacking ATP cones (Rose *et al.*, 2019).

Another type of ATP-cone was identified through bioinformatics, biochemical and structural studies of the class I RNR from *P. aeruginosa* (PaNrdA), which belongs to the subfamily NrdAz having multiple ATP-cones (Johansson *et al.*, 2016). Remarkably, in *P. aeruginosa* the ATP-cone domain was found to bind two molecules of dATP rather than the one molecule previously observed for the subfamily NrdAg (Jonna *et al.*, 2015). Binding of dATP induces the formation of a ring-shaped α_4 tetramer consisting of two PaNrdA dimers. Furthermore, the contacts between ATP-cone domains that lead to tetramer formation involve a different area of the surface of the ATP-cone to those in subfamily NrdAg.

In some organisms, the ATP-cone is not found at the N-terminus of the catalytic subunit NrdA but rather at the N-terminus of the radical generating subunit NrdB (Rozman Grinberg *et al.*, 2018a). In these organisms the NrdA subunit lacks an ATP-cone. Biochemical and structural studies of the class I RNR from the marine bacterium *L. blandensis* have shown that the ATP-cone responds to the binding of dATP by inducing formation of an unusual β_4 tetramer of two NrdB dimers, held together only by the interaction of the ATP-cones (Rozman Grinberg *et al.*, 2018a). The details of these interactions are remarkably similar to those seen in PaNrdA despite the fact that the ATP-cone is fused to a subunit completely different in size and fold. The ATP-cone thus appears to be a genetically transposable "molecular adhesive" that can induce protein-protein interactions somewhat independently of the subunit it is attached to. The *L. blandensis* NrdB (LbNrdB) was also found to belong to a family of NrdBs with a novel dimanganese metal center that does not use a tyrosyl radical. A high-valent Mn_2(III,IV) radical is instead generated directly at the metal center, as shown both for LbNrdB the closely related *Facklamia ignava* NrdB (FiNrdB) (Rozman Grinberg *et al.*, 2018a; Rozman Grinberg *et al.*, 2019).

When an ATP-cone-containing β is co-incubated with α, inactive oligomers with stoichiometries $\alpha_2\beta_4$ and $\alpha_4\beta_4$ can be isolated using size exclusion chromatography in the presence of dATP, for both the *L. blandensis* system (Rozman Grinberg *et al.*, 2018a) and the one from *F. ignava* (Rozman Grinberg *et al.*, 2018b). These complexes are inactive, since activity assays demonstrated a K_i of 20 μM for dATP, and at $\sim$90 μM dATP the enzyme is fully inactive.

One important unresolved question is how the dATP-induced tetramerization of LbNrdB and NrdBs with similar ATP-cone fusions inhibits enzymatic activity, since in the dATP-induced LbNrdB tetramer, the face of NrdB that should interact with NrdA in the active $\alpha_2\beta_2$ NrdAB complex is left exposed (Rozman Grinberg *et al.*, 2018a). How then do the α_2 dimers bind to the tetramer in such a way that the resulting complexes are inactive? In 1994, Uhlin and Eklund proposed a symmetrical docking model for the active $\alpha_2\beta_2$ complex in *E. coli*, based on the assumption that the symmetry axes of the dimers would coincide and that there would be maximal alignment of residues in the long-range PCET pathway (Uhlin and Eklund, 1994). The LbNrdB tetramer would superficially seem to present no hindrance to formation of such an active complex. However, a recent structure of the active *E. coli* RNR holoenzyme (EcNrdAB) determined using cryo-electron microscopy showed that the active complex was in fact highly asymmetrical (Kang *et al.*, 2020). One EcNrdA monomer makes extensive interactions with one EcNrdB monomer, in which large stretches of the C-termini of EcNrdA and EcNrdB become ordered in the interface, while the other pair of monomers make almost no interactions (Figure 1, lower left). One hypothesis as to how the tetramerization of LbNrdB via the ATP-cone inhibits enzymatic activity is by sterically blocking the formation of such an asymmetric complex. Thus one important question is the structural organisation of the inactive complexes of this class of RNRs and how they differ from those of active complexes.

Here we present a study of the low-resolution three-dimensional structures of the two components of *L. blandensis* RNR in the presence of the inhibitory nucleotide dATP, namely LbNrdA alone, LbNrdB alone, and the dATP-induced inactive RNR complex LbNrdAB with stoichiometry $\alpha_4\beta_4$, using small-angle X-ray scattering and cryo-electron microscopy. The results reveal a quasi-symmetrical complex in which the dimer axes of the LbNrdA and LbNrdB dimers are almost aligned and are consistent with a model in which the ATP cone sterically prevents the formation of a more asymmetrical active complex.

Materials and methods

LbNrdA homology model preparation

A homology model for residues 40–596 of LbNrdA was made using the SwissModel server (Waterhouse *et al.*, 2018) and the sequence of LbNrdA from UniProt (accession no. A3XHF8, gene MED217_17,130). The best template found was the 1.76 Å resolution crystal structure of the class Id NrdA from *Actinobacillus ureae* (PDB ID 6DQX) (Rose *et al.*, 2019), which has 59% sequence identity to LbNrdA over 92% of the length of LbNrdA. In addition, models were generated using the Phyre2 (Kelley *et al.*, 2015) and I-TASSER (Yang *et al.*, 2015) servers. Geometry statistics for all models were evaluated using the SwissModel server, including analysis using MolProbity (Chen *et al.*, 2010) and are presented in Supplementary Table 1, along with RMS deviations in Cα positions between the models.

LbNrdA/LbNrdB complex model preparation

The LbNrdA homology model obtained from SwissModel was placed on both sides of the LbNrdB tetramer complex (PDB ID 5OLK) in orientations compatible with the docking model for active class Ia RNR from *E. coli* proposed in 1994 (Uhlin and Eklund, 1994) by superimposing one of the dimers of 5OLK with the dimer of *E. coli* NrdB in the docking model, which was a gift from Ulla Uhlin and Hans Eklund.

Expression and purification of LbNrdA and LbNrdB

Both LbNrdA and LbNrdB were purified as described previously (Rozman Grinberg *et al.*, 2018a), except that the 6-His tags on the N-termini of both proteins were not removed. Protein concentrations were determined by A_{280} measurements using a Nanodrop spectrophotometer (Thermo Scientific) using molar extinction coefficients of 91,135 and 46,870 M^{-1} cm^{-1} for LbNrdA and LbNrdB respectively, as calculated from ProtParam (Gasteiger *et al.*, 2005). Further preparation of the protein samples for SAXS experiments is described below.

SAXS data collection

SAXS experiments were carried out at beamline BM29 of the ESRF, Grenoble, France (Pernot *et al.*, 2013) at a wavelength of 0.9919 Å. Data for LbNrdA with dATP were collected in batch mode. Dilution series of 5, 2.5, 1.25, and 0.625 mg/ml were prepared in buffer containing 50 mM Tris-HCl pH 7.6, 300 mM NaCl, 2 mM DTT, 5% glycerol, 2 mM magnesium acetate and 1 mM dATP, incubated for 5 min and centrifuged at high speed. The same buffer was used for buffer subtraction. Samples were loaded using the ESRF/EMBL SAXS sample changer (Round *et al.*, 2015). Software implemented at the beamline was used for data collection, radial averaging of the images, background subtraction and conversion to 1D scattering profiles (Supplementary Figure 1). The protein solutions were flowed slowly through the quartz SAXS measurement capillary (diameter 1.5 mm) to avoid radiation damage and the temperature was kept constant at 10°C. The scattered intensities were recorded on a pixel detector, Pilatus 1 M (Dectris), at a sample-detector distance of 2.867 m, allowing a range of momentum transfer q from 0.025 to 5.0 nm^{-1} ($q = 4\pi \sin \Theta/\lambda$, where 2q is the scattering angle and λ is the X-ray wavelength). Twenty frames of 45 ms each were recorded and compared to each other to check for radiation damage before merging. Further details are provided in Table 1.

Table 1: SAXS data collection and derived parameters. *Ab initio* models were calculated in expert mode where the number of knots used from the scattering curve was adjusted to make better agreement between the data and model. Structural parameters from crystal structures or homology model were calculated using CRYSOL.

Data collection parameters	**LbNrdA + LbNrdB dATP**	**LbNrdA + dATP**	**LbNrdB + dATP**
Experimental q range (nm^{-1})	0.0208–4.941	0.0208–4.941	0.0208–4.941
Exposure time (s)	0.045	0.045	0.045
Protein concentrations (mg ml^{-1})	n/a (SEC)	0.62, 1.25, 2.5, 5.0	n/a (SEC)
Structural parameters			
$I(0)$ [from Guinier/$p(r)$]	298.86/302.20	79.51/79.51	176.06/175.6
R_g (nm) [from Guinier/$p(r)$]	6.36 ± 0.11/ 6.42 ± 0.11	3.84 ± 0.04/ 3.88 ± 0.02	4.51 ± 0.06/ 4.49 ± 0.03
Shell R_g (nm) from crystal structure or model	7.31	4.56	4.81
D_{max} (nm)	22.6 ± 0.2	13.37 ± 0.02	15.1 ± 0.1
Envelope diameter (nm) from crystal structure or model	19.4	12.2	13.2
Porod volume estimate (nm^3)	724.06	221.93	375.67
Molecular mass determination			
Average excluded volume, V_{ex} ($Å^3$)	726.159	210.832	378.044
Calculated monomeric M_r from sequence (kDa)	α: 70.66, β: 51.78 $\alpha_4\beta_4$: 489.76 $\alpha_2\beta_4$: 348.44	α: 70.66, α_2: 141.32	β: 51.78, β_4: 207.12
Mol. Mass from excluded volume (kDa)	427.15	124.02	222.38
Mol. Mass from porod volume (kDa)	425.91	130.55	220.98
Mol. Mass from SAXSMoW	466.5	141.7	254.4
Modelling parameters			
Ab initio analysis	*GASBOR*	*GASBOR*	*DAMMIN*
Symmetry imposed	P222	P2	P222
Number of knots used	58	74	86
Validation and averaging	*DAMAVER*	*DAMAVER*	*DAMAVER*
χ^2	1.38 ± 0.233	1.22 ± 0.125	0.99 ± 0.007
NSD value	1.035 ± 0.013	1.097 ± 0.035	0.726 ± 0.059
DENSS estimated resolution by FSC = 0.5 criterion (Å)	49.7	28.2	41.5

For the complex of LbNrdA and LbNrdB, as well as LbNrdB with dATP, in-line size exclusion chromatography with SAXS (SEC-SAXS) was employed, where purified protein sample was loaded onto a HiLoad 10/300 GL Superdex S200 (GE Healthcare) column. For the complex, NrdA and NrdB were mixed to reach a final concentration of 85 μM each, in buffer containing 50 mM Tris-HCl pH 7.6, 300 mM NaCl, 10% glycerol, 2 mM DTT, 1 mM dATP and 2 mM magnesium acetate. After 5 min incubation, the mixture was centrifuged at high speed and 350 μL were injected onto the column. The proteins were eluted at a flow rate of 0.5 ml min^{-1} and passed through the capillary cell. Frames were collected every 1 s with a total of 3,000 frames. The buffer for SEC-SAXS was 50 mM Tris-HCl pH 7.6, 300 mM NaCl, 2 mM DTT, 1 mM Mg acetate, 0.4 mM dATP and 5% glycerol. LbNrdB with dATP was prepared in the same buffer as the LbNrdA-LbNrdB complex, 350 μL of 11 mg/ml protein were loaded into the column and the same flow rate and SEC buffer were used in the SEC-SAXS data collection. Total scattering profiles and Rg plots for the elutions are shown in Supplementary Figures 2, 3.

SAXS data processing

The datasets collected in batch mode at different protein concentrations were compared to look for concentration-dependent aggregation effects. Having found no such artefacts, the datasets were then processed and scaled using Primus (Konarev *et al.*, 2003). The radii of gyration (R_g) of the proteins in solution were determined from the lowest q values of the SAXS data, using the Guinier approximation. The value I(0)/c, the scattering intensity at zero angle normalized to the protein concentration of the sample, is proportional to the molecular mass M of the protein, estimated after calibration of the intensity using reference samples. The distance distribution function, representing the distribution of distances between any pair of volume elements within the particle, together with the structural parameters derived from $p(r)$, i.e. the maximum dimension of the particle ($D_{\max}$), and the radius of gyration, were evaluated using the indirect transform method as implemented in the program GNOM (Guinier, 1939; Petoukhov *et al.*, 2012). SEC-SAXS profiles were processed by CHROMIXS (Panjkovich and Svergun, 2018) and best fractions were chosen according to their expected elution volumes based on their molecular weight. The figures of envelopes from SAXS data and their corresponding molecular models were made using the programs PyMOL (www.pymol.org), Situs (Wriggers, 2012) and Chimera (Pettersen *et al.*, 2004).

Ab initio SAXS modelling

The model calculations were performed using GASBOR (Svergun *et al.*, 2001) or DAMMIN (Franke and Svergun, 2009). The scattering profiles up to $q_{max} = 1.26$ to $4.2\,nm^{-1}$ were used and 20 independent runs of modelling were carried out with each method. The 20 models were averaged and filtered using DAMAVER (Volkov and Svergun, 2003). As an additional check on the oligomeric state of LbNrdA and LbNrdB and their complexes by GASBOR or DAMMIN, several *ab initio* dummy residue reconstructions were performed assuming monomeric as well as other oligomeric organizations. In all cases the models were generated assuming that either symmetry (P2 or P222) or no symmetry was present, and the best results were selected.

Electron density map generation using DENSS

As an additional check of the correctness of the *ab initio* structures obtained using GASBOR and DAMMIN, electron density maps were generated using the phase retrieval method implemented in the program DENSS (Grant, 2018). The same output pair distribution functions and D_{max} values from GNOM as used for DAMMIN and GASBOR were used. Twenty independent models were generated for each dataset. When appropriate, 2-fold symmetry was applied to the simulation (P222 is not possible in DENSS). The best principal axis of the ellipsoid describing the molecular dimensions to which to apply the symmetry was determined by applying it to each axis in turn in independent simulations.

Cryo-electron microscopy

Sample preparation

For making the LbNrdAB complex, 230 μL of LbNrdB were taken from an initial stock solution of 6 mg/ml, 2 mM dATP was added followed by 170 μL of LbNrdA from an initial stock solution of 10 mg/ml. The mixture was incubated for 10 min on ice. This was followed by centrifugation of the sample at 5,000 rpm for 6 min. Protein was initially present in buffer containing 50 mM Tris (pH 7.6), 50 mM NaCl, 2% glycerol, 2 mM $MgCl_2$, 2 mM TCEP and 0.4 mM dATP. A Superdex 200 column was used for the purification of the complex. Gentle glutaraldehyde crosslinking was used, as initial trials showed that the complex fell apart almost completely on cryo-EM grids. The original buffer was changed from 50 mM Tris to 50 mM HEPES as the Tris would react with the glutaraldehyde. Buffer exchange was done on the complex using an Amicon Ultra 4 ml centrifugal filter membrane. All other buffer components remained the same. Following wash of the Superdex 200 column using HEPES buffer, 4 ml of 0.05% glutaraldehyde were injected into the column. A further 2 ml of HEPES buffer (without glutaraldehyde) were then injected, followed by injection of the protein complex and elution. Fractions of 200 μL were collected. The formation of $\alpha_4\beta_4$ complexes (theoretical molecular weight 490 kDa) was verified by the presence of a band at >460 kDa on a Coomassie-stained reducing SDS-PAGE gel (not shown).

Single particle cryo-EM grids were prepared using Quantifoil R 1.2/1.3 grids (without continuous carbon coating) using sample concentration of 2 mg/ml. The Vitrobot settings were T = 4°C, humidity 100%, blot time 5 s, blot force −5, wait and drain time 1 and 0 s respectively. The grids had a good particle distribution in thin ice.

Data collection

Data were collected from the grids using a Titan Krios microscope (ThermoFisher) equipped with a Gatan K2 detector. A total of 493 movies were collected in counting mode using an accelerating voltage of 300 kV. The data acquisition parameters were as follows: objective aperture 100, energy filter slit 20 eV, illuminated area 1.02 μm, spot size 6 nm, spherical aberration 2.7, defocus range -1.5 to -1.3 μm, pixel size 0.82 $Å^2$, dose 11.77 $e^- Å^{-2} s^{-1}$ with an exposure time of 4 s, making a total dose of 47.1 $e^- Å^{-2}$. The total number of dose fractions was 40.

Data processing

Data processing was done using cryoSPARC v2.15 (Punjani *et al.*, 2017; Punjani, 2020). A total of 493 movies were imported then patch motion correction (multi) and patch CTF estimation (multi) were carried out using default parameters. A volume of a hypothetical octamer was made with one pair of LbNrdA dimers from the homology model and one LbNrdB tetramer from the crystal structure, based on the docking model of active complex suggested by Uhlin and Eklund, and was imported into cryoSPARC. From this volume, a set of 50 equally spaced templates was created for template-based particle picking from the CTF-corrected micrographs. The particle diameter was 210 Å and the minimum separation distance was 0.5 particle diameters. This template, low-pass filtered at 20 Å, gave 200,948 particles from the 493 micrographs. Particles with an NCC score <0.300 and power threshold <961 and $>4{,}971$ were then discarded, leaving 149,794 particles, which were then extracted with a box size of 448 pixels. All particles were used for 2D classification with 80 classes and default parameters. Fifteen 2D classes were selected. The 2D classes showed the tetramer of LbNrdB and one dimer of LbNrdA distinctly while the other dimer of LbNrdA seemed less well ordered. These 2D classes had 35,473 particles, which were used for a second round of 2D classification into 50 classes. From here, 27 2D classes having 29,902 particles were selected and used in two separate runs to make a single *ab initio* model and two *ab initio* models. The particles from the single model were then refined against the two models using heterogeneous refinement. One of the resulting models (model A) had 19,630 particles and a gold standard FSC resolution of 9.9 Å while the other had 10,272 particles with FSC resolution

10.8 Å. Finally, the 19,630 particles were refined against model A using homogeneous refinement, which produced a refined model with FSC resolution 8.2 Å.

Alignment of crystal structures and homology models to *ab initio* envelopes from SAXS

The bead models from DAMMIN and GASBOR, which were obtained from independent 20 averaged and filtered models, were converted to volumes using SITUS. The crystal structures and homology models were aligned to the resulting envelopes using a combination of rigid body fitting in Chimera and manual adjustment. Molecular structures were aligned to the DENSS volumes using the script denss.align.py and some manual adjustment.

Results

LbNrdA (α_2) *homology model*

The LbNrdA construct used consists of 600 amino acids. No crystal structure of LbNrdA has been determined to date. A reliable homology model for residues 40–596 was prepared using the SwissModel server with the dimeric crystal structure of *Actinobacillus ureae* (PDB ID 6DQX) (Rose *et al.*, 2019) as template. This NrdA has 59% sequence identity to LbNrdA over 92% of the length of the latter. The QMEAN score for the homology model was −1.20. Two further models were constructed using the Phyre2 and I-TASSER servers. All three models were evaluated using the same criteria on the SwissModel server. Comparative statistics are shown in Supplementary Table 1. The Phyre2 and I-TASSER models had poorer overall quality but predicted a structure for the N-terminal 39 residues based on weak homology to three helices in the "connector domain" (Parker *et al.*, 2018; Rose *et al.*, 2019) that falls between the ATP-cone and the core in many class Ia RNRs. However, this three-helix bundle packed against the core of the protein such that its hydrophobic core enclosed three charged residues from the surface of the core: Glu69, Arg72, and Asp76. Furthermore, the predicted

3-helix bundle clearly protrudes from the cryo-EM reconstruction (see below). No homology to known structures could be found for the first 39 residues of LbNrdA using a BLAST search (Altschul *et al.*, 1990). Taken together, the evidence points towards the N-terminal 39 residues of LbNrdA being disordered.

As expected, the homology model strongly resembles an NrdA dimer lacking an ATP-cone. The SwissModel coordinates were used for further interpretation of the interactions between LbNrdA and LbNrdB. The crystal structure with PDB ID 5OLK was used for NrdB (Rozman Grinberg *et al.*, 2018a).

Small-angle X-ray scattering

SAXS data were collected for three samples: LbNrdA in the presence of dATP, LbNrdB in the presence of dATP, and finally the inactive complex of LbNrdA and LbNrdB in the presence of dATP (LbNrdAB). To ensure sample homogeneity, the full-length LbNrdB and LbNrdAB preparations were recorded using inline size exclusion chromatography. Key structural parameters derived from the SAXS data are presented in Table 1.

LbNrdA forms a dimer in the presence of dATP

The solution scattering data from dATP-bound LbNrdA indicate a globular protein with an R_g of 3.92 nm and D_{max} of 13.9 nm (Table 1). The R_g value compares well with the value of 3.66 nm from the homology model, which lacks a total of 43 residues at the termini of each monomer, consistent with its lower R_g. The distance distribution function is shown in Figure 2. The form of the $p(r)$ function, with a maximum at less than $D_{max}/2$, suggests that dATP-bound LbNrdA NrdA forms a species that is wider in one axis and the long tail on the right is consistent with the presence of unstructured terminal residues.

Furthermore, models for dATP-bound LbNrdA obtained by GASBOR *ab initio* reconstruction and DENSS phase recovery show a good fit to the dimeric homology model (Figure 2). The resolution

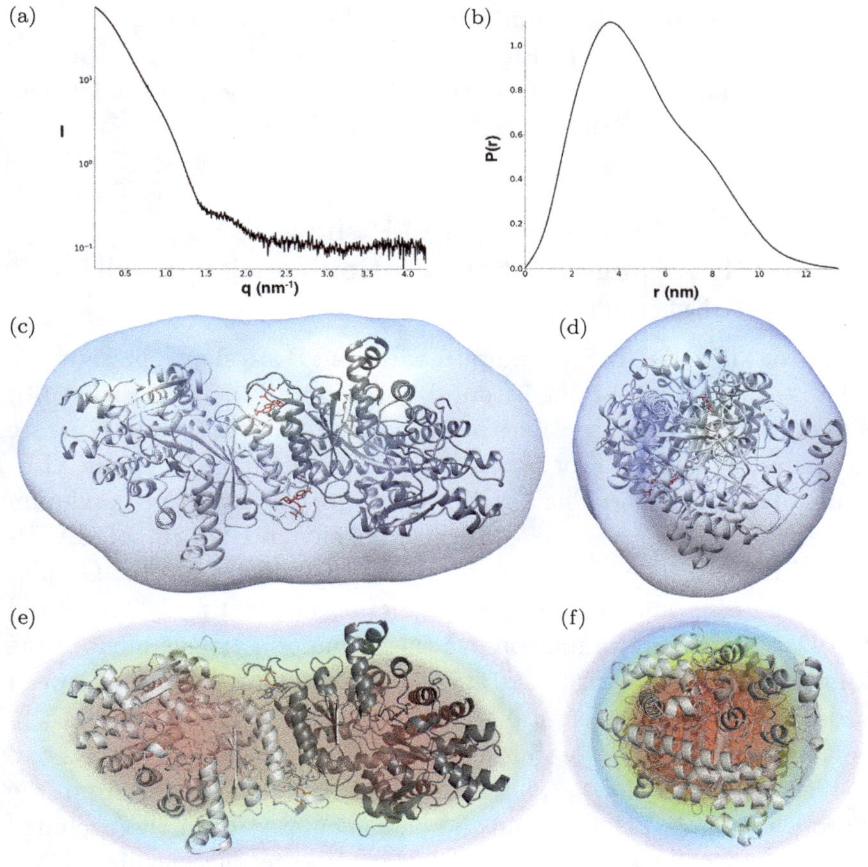

Figure 2: **(a)** Scattering curve (black line) and fitting (red line) to generate pair distance distribution function of dATP-bound LbNrdA. **(b)** Pair distance distribution function. The angular range used for fitting to generate $p(r)$ is from 0.13 to 4.22 nm. *Ab initio* models of dATP bound LbNrdA superimposed on the homology model. **(c, d)** Two views of the envelope from GASBOR processed to a volume using SITUS, related by a 90° rotation around the vertical axis. **(e, f)** Similar views of the DENSS density reconstruction at approximately 42 Å resolution.

of the DENSS reconstruction is estimated to be 28 Å by the FSC 0.5 criterion (Supplementary Figure 4). The correlation score of a 15 Å resolution map calculated from the homology model and the DENSS reconstruction is 0.889, as calculated in DENSS.

LbNrdB in complex with dATP shows a tetrameric arrangement in solution consistent with the crystal structure

SAXS data for LbNrdB in complex with dATP are shown in Figure 3. The $p(r)$ function obtained from dATP-bound LbNrdB (Figure 3b) shows tailing at high values of r consistent with small separate domains connected to the ends of the core domain, i.e. the four ATP-cones of the tetramer. The distribution of $p(r)$ peaks with a maximum at less than $D_{\mathrm{max}}/2$ also confirms the elongated form of dATP-bound LbNrdB. The $\mathrm{R_g}$ values of 4.49 nm obtained from the Guinier approximation and from the $p(r)$ function are identical. These and the D_{max} value of 15.1 nm confirm the tetrameric arrangement in the known crystal structure (5OLK). The maximum distance between ordered atoms in the crystal structure is about 12.5 nm, thus D_{max} is consistent with the presence of disordered residues at the N- and C-termini (see below). The molecular mass calculated from the Porod volume is 221 kDa and the one calculated from the excluded volume is 222 kDa, both of which are close to the calculated one of 250 kDa, confirming the tetrameric arrangement of LbNrdB in solution.

The overall *ab initio* models obtained from the SAXS data also show good agreement with the crystal structure. The resolution of the DENSS reconstruction is estimated to be 42 Å (Supplementary Figure 4). The correlation score between the DENSS reconstruction and a 15 Å map calculated from the crystal structure is 0.740. The excluded volume obtained from the model is also consistent with a homotetrameric oligomeric state. The crystal structure of LbNrdB lacks between 24 and 29 residues at the N-terminus and 28–29

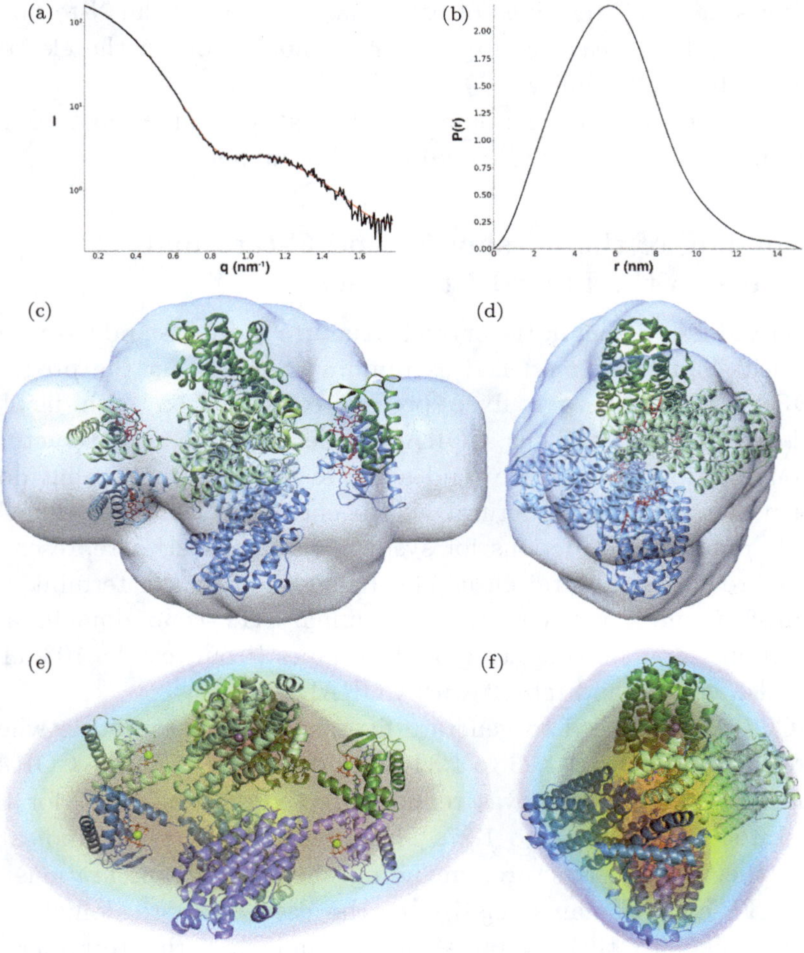

Figure 3: **(a)** Fit of the GASBOR model (red line) to the experimental scattering curve (black line); **(b)** pair distance distribution function of LbNrdB in the presence of dATP. The angular range used for fitting to generate $p(r)$ was from 0.13 to 1.77 nm^{-1}. **(c–f)** The tetrameric arrangement of LbNrdB as in PDB structure 5OLK fits well with the *ab initio* models derived from solution scattering data. **(c, d)** show the crystal structure fitted to the reconstruction from DAMMIN processed to a volume in SITUS. **(e, f)** show the fit of the crystal structure to the electron density reconstructed by DENSS at approximately 28 Å resolution. Views in **(d, f)** are rotated by 90° around the vertical axis relative to **(c, e)** respectively. The cartoon representation of LbNrdB is coloured by chain. The 8 dATP molecules are shown as sticks and the 4 Mg^{2+} ions as spheres.

residues at the C-terminus of each chain, including the N-terminal His-tag and its cleavage site, which are not visible in the electron density due to flexibility.

These regions may account for the extra volume close to the N-terminus (near the ATP-cone) and C-terminus.

Modelling of the missing N- and C-terminal residues in the LbNrdB tetramer

The residues missing in the crystal structure at the N- and C-termini of each chain were modelled as dummy atoms using the program CORAL, which uses a fixed protein conformation with flexible ends and/or linkers. Thus, CORAL modelling gives good structural reconstructions for a more rigid and less flexible protein that does not possess a large structural heterogeneity. In contrast, CORAL yields poor reconstructions for systems that are better represented by a broader structural ensemble. In addition to the termini, the connecting neck between the N-terminal ATP-cone domain and C terminal domain consisting of 16 residues (from residue 103–118) was also deleted and rebuilt with CORAL.

CORAL was used to generate five representative models where χ^2 values varied from 1.23 to 1.32 (Figure 4a). Through the CORAL modelling the χ^2 value was reduced from on average 3.27 for the crystal structure alone to 1.27. This good fit (Figure 4b) indicates that the overall structure, including the relative orientations of the ATP-cone domains relative to the catalytic core domains, is very similar to the crystal structure and that the tetramer as a whole is not very flexible. In the CORAL model, in all four monomers the N-terminal residues project directly outwards from their respective ATP-cones. In contrast, the missing C-terminal residues show more variability in that they either lie close to the structured domains or project outwards in opposite directions. The modelled termini correspond well to the regions of density not occupied by the crystal structure in the DAMMIN and DENSS reconstructions.

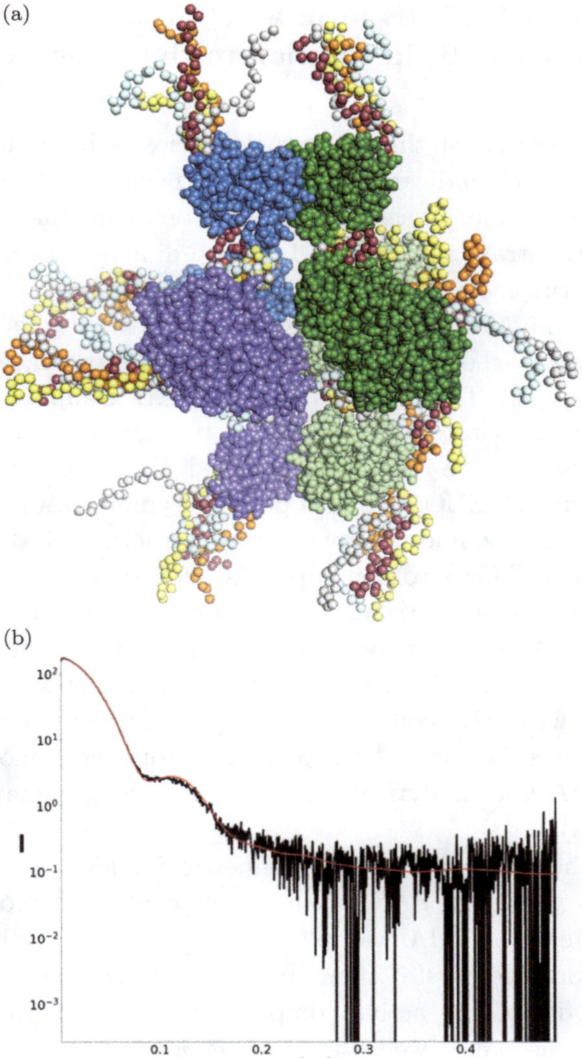

Figure 4: **(a)** Models of the LbNrdB tetramer with built missing N- and C-termini, each with 24–29 residues modelled using CORAL. The neck region between the core and ATP cone domains consisting of 16 residues was also deleted and rebuilt. Five possible models obtained from independent runs of CORAL are shown. **(b)** The scattering curve of LbNrdB (black) and its fit with the CORAL model (red).

LbNrdA and LbNrdB form a hetero-octamer with stoichiometry $\alpha_4\beta_4$ in solution in the presence of dATP

The Porod volume of the LbNrdAB species induced by dATP is 724.1 nm^3, which indicates a molecular mass of 425.9 kDa. The calculated molecular mass of an $\alpha_4\beta_4$ octamer that consists of the LbNrdB tetramer with two LbNrdA dimers attached is 489.8 kDa, while the molecular mass of an $\alpha_2\beta_4$ hexamer is only 348.4 kDa. Thus the most plausible stoichiometry for this species in solution is $\alpha_4\beta_4$, though we cannot exclude the presence of a small amount of $\alpha_2\beta_4$. The R_g of the LbNrdAB complex is 6.26 nm and D_{max} is 22.6 nm, both of which are significantly larger than either of the β_4 or α_2 species described above. During *ab initio* modelling with GASBOR, several possible symmetries were imposed and only P222 symmetry gave a reconstruction that fitted well with the data. This strongly supports an arrangement where two dimers of NrdA are arranged on opposite sides of the dATP-induced NrdB tetramer observed in the crystal structure and confirmed by SAXS (Figure 5). Furthermore, this arrangement is compatible with the concentration-dependent formation of $\alpha_4\beta_4$ species observed in previous gas phase electrophoretic mobility analysis GEMMA and analytical SEC experiments (Rozman Grinberg *et al.*, 2018a).

To test the compatibility of a symmetrical model for the inhibited $\alpha_4\beta_4$ complex with the SAXS data, one copy of the homology model for the dimeric LbNrdA was placed on each side of the LbNrdB tetramer from the crystal structure in a way compatible with the docking model of the active complex of *E. coli* (Figure 5). This model fits reasonably well with the *ab initio* models, both from GASBOR and DENSS. The resolution of the DENSS model was estimated to be 50 Å (Supplementary Figure 4). The correlation score of a 15 Å map of the docking model with the DENSS volume was 0.768 as calculated by DENSS. Once again, the extra density present at both ends is most likely explained by the flexible N-terminus.

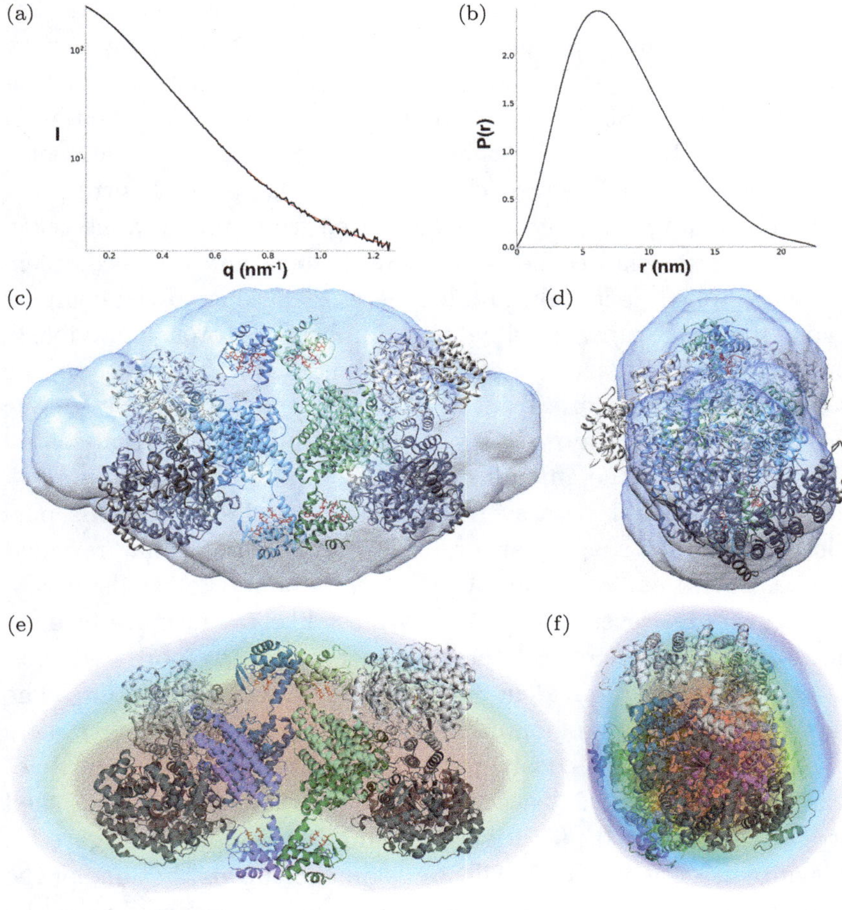

Figure 5: **(a)** Scattering curve of the LbNrdA/NrdB complex in the presence of dATP. **(b)** $p(r)$ function. The angular range used to generate $p(r)$ is from 0.11 to 1.26 nm. **(c, d)** Approximately orthogonal views of a fit of the docking model of the complex to the envelope from GASBOR. **(e, f)** Approximately orthogonal views of the fit of the same model to the DENNS reconstruction.

Cryo-EM of the dATP-inactivated LbNrdAB complex

Analytical SEC had previously demonstrated the existence of species with molecular weights consistent with both $\alpha_4\beta_4$ and

$\alpha_2\beta_4$ stoichiometries, with the stoichiometry being concentration-dependent (Rozman Grinberg *et al.*, 2018a). Initial attempts were made to prepare cryo-EM grids of the LbNrdAB complex by isolating fractions from a SEC run where LbNrdA had been mixed with excess LbNrdB. However, despite taking fractions from the centre of the peak in a well-defined elution profile that should correspond to the $\alpha_4\beta_4$ complex, 2D classes generated from micrographs taken from these grids showed near-complete dissociation of the complex. We thus applied gentle glutaraldehyde cross-linking to the complex during the SEC run according to a published protocol previously applied to a GPCR/arrestin complex (Shukla *et al.*, 2014). Here, the protein travels through a front of glutaraldehyde as it traverses the SEC column. Furthermore, the sample was applied to carbon-coated grids to maximize the amount of intact complex.

Two-dimensional class averages from a small set of 29,902 particles (Figure 6a) suggest an organization similar to that revealed by SAXS, with one α_2 dimer symmetrically bound to each side of a β_4 tetramer. However, the sample is most likely a mixture of $\alpha_4\beta_4$ and $\alpha_2\beta_4$ species, as one of the α_2 dimers is less well-ordered than the other, as already evident from the 2D classes (Figure 6a). The grids also contained a population of non-complexed β_4 tetramers (not shown). A 3D reconstruction was made from these particles with a resolution of 8.2 Å according to a gold standard Fourier shell correlation value of 0.143 (Figure 6b and Supplementary Figure 5). No symmetry was imposed. The 3D volume confirms that one of the α_2 dimers is significantly less well ordered or less occupied than the other, as the density is discontinuous and appears at a lower contour level than for the more ordered dimer (right hand side of Figure 6b). The dimeric homology model from the present work was fitted as a rigid body into the volume for α_2 using Chimera and the crystal structure of the β_4 tetramer was fitted into the corresponding volume as a single rigid body. Considering the more well-ordered of the two α_2 dimers, it is arranged on top of one of the dimers of the β_4 tetramer such that the active sites of both α_2 monomers face the β subunits. However, the 2-fold axes of α_2 and β_2 are not exactly superimposed and the α_2 dimer is tilted such that one of the monomers approaches

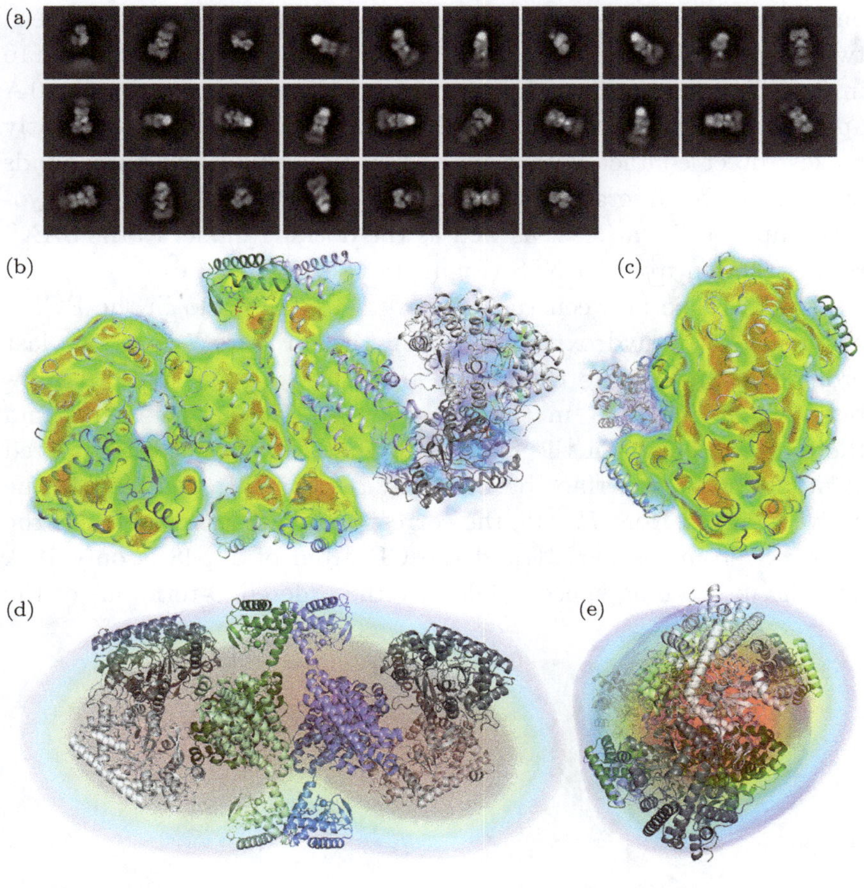

Figure 6: **(a)** Initial 2D classification of 29,902 particles of the cross-linked LbNrdAB complex in cryoSPARC. The first 27 classes are shown. **(b, c)** Approximately orthogonal views of the cryo-EM volume of the $\alpha_4\beta_4$ complex at 8.2 Å α_2 dimers are shown in two shades of grey while the two dimers of the β_4 tetramer are shown in shades of green and blue respectively. The volume is coloured according to electrostatic potential level, from transparent blue at 2 σ through cyan at 6 σ, green at 8 σ, yellow at 10 σ to red at 15 σ, where σ is the number of standard deviations of the potential above the mean for the reconstruction box. Note the strong density for several helices. The lack of potential at higher levels for the right-hand dimer indicates lower occupancy and/or higher disorder. **(d, e)** Approximately orthogonal views of the fit of the cryo-EM model to the SAXS envelope from DENSS (the GASBOR envelope is not shown for brevity).

one of the ATP-cones more closely. The shortest distance between two Cα atoms on one side is ~10 Å, between Lys560 on α and Thr16 in β, while in the other pair of monomers the same atoms are ~20 Å apart. Interestingly, neither of the ATP-cone domains apparently makes direct contact with the α_2 dimers through hydrogen bonds or hydrophobic interactions. The molecular model fitted to the cryo-EM map also fits at least as well as the docking model to the SAXS reconstruction from DENSS (Figures 6d, e).

The distance between the last ordered amino acid in the PCET chain in α (the hydroxyl group of Tyr579; Figure 7) and the last ordered residue of the chain in β (the Cβ atom of Trp133) varies between 26 and 29 Å in the four pairs of interacting chains and there is no density in the 8.2 Å map that clearly suggests ordered structure at the interface between any α/β pair. In contrast, in the active complex from *E. coli*, the corresponding distance between the hydroxyl group of Tyr731 and the CB atom of Trp48 is only 19 Å and the intervening space is filled by the ordered C-terminus of the

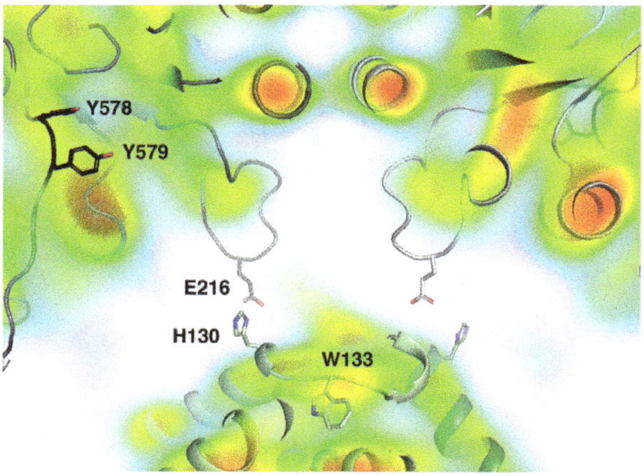

Figure 7: Closeup view of the central part of the interaction area between LbNrdA and LbNrdB in the dATP-inhibited complex, closest to the misaligned 2-fold symmetry axes. The homology model of LbNrdA and the crystal structure of LbNrdB are superposed on the 8.2 Å cryo-EM reconstruction coloured as in Figure 6. Note the lack of electrostatic potential in the interface between the two proteins. The potential area of closest contact involves the loop containing Glu216 in the α subunit.

β subunit, including Tyr356, which is critical for the PCET pathway (Kang *et al.*, 2020).

The closest approach between the α and β subunits occurs between the tip of the loop 213–222 between β-strands C and D of the α subunit and helix 118–131 on the β subunit (Figure 7). The distance between the Cα atom of Asp α216 and His β130 is around 10 Å. This interaction area is close to Trp133, one of the last residues in the PCET pathway on β. The side chains of α216 and β130 may be projected towards each other, but this interaction would be insufficient to stabilize the formation of a complex. It should be borne in mind that the α subunit in the current analysis is a homology model and that the interaction between subunits may involve structural elements not resolved at the current resolution. There is no well-defined density for this loop in the cryo-EM reconstruction. Work is ongoing to deconvolute possible heterogeneity in the particle population and to obtain a larger dataset for higher resolution.

Discussion

Class I ribonucleotide reductases are dependent for their activity on the formation of a productive complex between the α and β subunits to enable long-range proton-coupled electron transfer from β to α and back again on each catalytic cycle. The RNRs from *L. blandensis* and *F. ignava* are both inactivated by the formation of a tetramer of β subunits in the presence of dATP that is catalytically incompetent, most likely by holding the subunits in a relative orientation that breaks the PCET chain, as has been observed previously in various forms for other class I RNRs (Ando *et al.*, 2011; Fairman *et al.*, 2011). Here we have studied the oligomeric organization of the individual α and β components from *L. blandensis*, as well as the inactive complex, all in the presence of the inhibitory nucleotide dATP. The SAXS- derived structures agree well with a homology model for the α_2 dimer and the previously published crystal structure of the β_4 tetramer. Furthermore, the SAXS studies suggest that the ATP-cones of the β_4 tetramer are not highly flexible in solution with respect to the core domains, and good agreement to the experimental scattering profile can be obtained simply by modelling the additional disordered residues at the N- and C-termini. Thus, the crystal structure is a faithful representation of the conformation in solution.

Biochemical and biophysical studies of the dATP-inactivated complex indicated the concentration-dependent formation of oligomers with stoichiometry $\alpha_2\beta_4$ and $\alpha_4\beta_4$ when α_2 dimers were added to β_4 tetramers (Rozman Grinberg *et al.*, 2018a). However, the structures of these complexes, and thus the molecular mechanisms of inactivation of class II RNRs, in which the ATP-cone is found in the β subunit, have not been elucidated previously. Here we studied these complexes with both SEC-SAXS and cryo-EM. The SAXS studies are consistent with a predominantly $\alpha_4\beta_4$ stoichiometry and a hypothetical structure in which two α_2 dimers attach symmetrically or pseudo-symmetrically to each side of a β_4 tetramer. Such an arrangement would be inconsistent with an intact PCET pathway, as cryo-EM studies on the class I RNR from *E. coli* have indicated that such a complex is in fact highly asymmetrical (Figure 1). However, the SAXS envelopes are of too low resolution to dissect the structure of the complex in detail.

To confirm and extend the solution results, we also studied the inactivated complexes by single particle cryo-EM. The complexes were isolated using size exclusion chromatography in the presence of dATP using the same protocol as for SEC-SAXS. In addition, gentle on-column crosslinking using glutaraldehyde was applied. The latter was necessary because the forces exerted at the air-water interface in the cryo-EM freezing entirely disrupted the complexes without crosslinking, though the SEC-SAXS results show that the same complexes are stable in solution. A reconstruction was obtained at 8.2 Å resolution that also reveals binding of two α_2 dimers, one on either side of the β_4 tetramer, with both active sites in the α_2 dimer facing towards the active site of one β_2 dimer, in a near-productive conformation. Despite cross-linking to prevent disruption of the complex during grid preparation, one of the dimers is less well-occupied than the other, but the SAXS results are most consistent with an $\alpha_4\beta_4$, stoichiometry in solution. Most surprisingly, the cryo-EM volume shows the subunits at some distance from each other and in a slightly asymmetrical orientation such that one ATP-cone on β_4 approaches an α subunit more closely than does the other from the same dimer of the tetramer.

Tetramerization may inhibit *L. blandensis* RNR by blocking formation of an asymmetric active complex

The observation that the dATP-inhibited $\alpha_4\beta_4$ LbNrdAB complex is quasi-symmetrical in combination with the recent discovery that the active complex of the *E. coli* enzyme is highly asymmetrical led us to hypothesize about how the inhibition might occur for *L. blandensis, F. ignava* and similar organisms. We used the homology model of LbNrdA in combination with the crystal structure of the dATP-inhibited LbNrdB tetramer and superimposed these on the cryo-EM structure of the active *E. coli* $\alpha_2\beta_2$ complex (EcNrdAB, PDB ID 6W4X (Kang *et al.*, 2020)) to check for possible steric hindrances to the formation of an active complex (Figure 8). More specifically, we superposed the "active" monomer of the NrdB component of EcNrdAB (light green) on one chain of the LbNrdB tetramer. Then we superposed the homology model of LbNrdA (light grey) on the "active" monomer of EcNrdA from EcNrdAB. This provides a current best model of the interactions between LbNrdA and LbNrdB in the putative active complex. It can clearly be seen that in this conformation the ATP-cone of the "inactive" LbNrdB monomer (dark green in Figure 8) would clash with the "inactive" LbNrdA monomer (dark grey). Since the LbNrdA dimer is relatively rigid, this clash would also prevent the interaction of the two "active" chains.

Thus, despite subtle differences, our SAXS and cryo-EM results are both compatible with the hypothesis that the ATP-cone functions as a steric block to formation of the highly asymmetrical complex that would be necessary for an intact PCET chain. These results begin to illuminate how the ATP-cone functions as a transposable functional module that, through oligomerization and steric hindrance, can inhibit RNR activity even when transposed from its usual place in the catalytic subunit to a new one in the radical generating subunit.

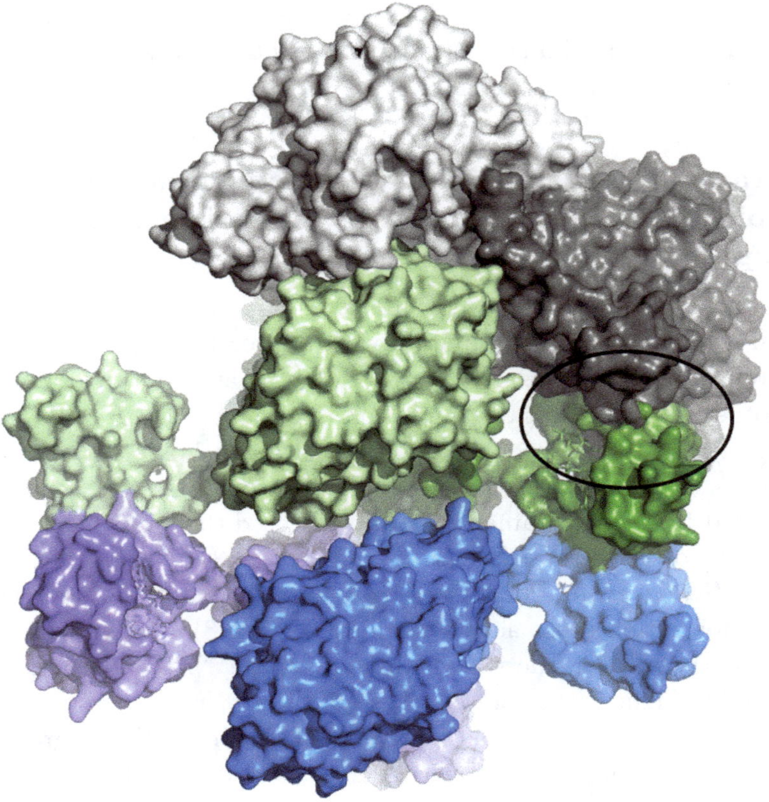

Figure 8: Model of the hypothetical steric clash that may prevent formation of an active complex of *L. blandensis* RNR when LbNrdB has tetramerized in the presence of dATP. Dimeric LbNrdA is shown in two shades of grey. The two dimers of LbNrdB that form the tetramer are shown in light and dark shades of green and blue respectively. The model has been generated to recreate the interactions that would occur in the active complex, based on the structure of *E. coli* NrdAB, between the light grey LbNrdA and pale green LbNrdB subunits. This orientation is prevented by a severe steric clash of the ATP-cone of the dark green LbNrdB subunit with the light grey LbNrdA subunit (circled).

Data availability statement

The cryo-EM data presented in this study can be found in the Electron Microscopy Database (EMDB) https://www.ebi.ac.uk/pdbe/emdb/, with accession number EMD-12985.

Author contributions

MH collected and analysed SAXS data. IB analysed SAXS data, prepared cryo-EM samples, collected and analysed the cryo-EM data. IRG prepared the LbNrdA and LbNrdB proteins used in this study, prepared samples for SAXS and collected the SAXS data. B-MS contributed to the discussion. DL conceived and supervised the experiments and analysis. All authors contributed to the manuscript.

Funding

This work was funded by grant 2016-04855 from the Swedish Research Council to DTL and 2019-01400 to BMS, as well as grant 2018/820 from the Swedish Cancer Foundation and a grant from the Wenner-Gren Foundations to BMS. IB is the recipient of a scholarship from the Lawski Foundation. Molecular graphics and analyses performed with UCSF Chimera, developed by the Resource for Biocomputing, Visualization, and Informatics at the University of California, San Francisco, with support from NIH P41-GM103311.

Acknowledgments

We thank Petra Pernot and other staff at beamline BM29 of the ESRF for help with SAXS data collection and Michael Hall for help with cryo-EM grid preparation and data collection. The cryo-EM data were collected at the Umeå Core Facility for Electron Microscopy, a node of the cryo-EM Swedish National Facility, funded by the Knut and Alice Wallenberg, Family Erling Persson and Kempe Foundations, SciLifeLab, Stockholm University and Umeå University.

The chapter was originally published in Frontiers in Molecular Biosciences, Hasan M, Banerjee I, Rozman Grinberg I, Sjöberg B-M and Logan DT (2021) Solution Structure of the dATP-Inactivated Class I Ribonucleotide Reductase From Leeuwenhoekiella blandensis by SAXS and Cryo-Electron Microscopy. *Front. Mol. Biosci.* 8:713608. doi:10.3389/fmolb.2021.713608.

Supplementary material

The Supplementary Material for this article can be found online at: https://www.frontiersin.org/articles/10.3389/fmolb.2021.713608/full#supplementary-material

References

Ahmad, M. F., and Dealwis, C. G. (2013). The Structural Basis for the Allosteric Regulation of Ribonucleotide Reductase. *Prog. Mol. Biol. Transl Sci.* 117, 389–410. doi:10.1016/B978-0-12-386931-9.00014-3

Altschul, S. F., Gish, W., Miller, W., Myers, E. W., and Lipman, D. J. (1990). *Basic Local Alignment Search Tool. J. Mol. Biol.* 215, 403–410. doi:10.1016/S0022-2836(05)80360-2S0022-2836(05)80360-2

Ando, N., Brignole, E. J., Zimanyi, C. M., Funk, M. A., Yokoyama, K., Asturias, F. J., *et al.* (2011). Structural Interconversions Modulate Activity of *E. coli* Ribonucleotide Reductase. *Proc. Natl. Acad. Sci.* 108, 21046–21051. doi:10.1073/pnas.1112715108

Ando, N., Li, H., Brignole, E. J., Thompson, S., McLaughlin, M. I., Page, J. E., *et al.* (2016). Allosteric Inhibition of Human Ribonucleotide Reductase by dATP Entails the Stabilization of a Hexamer. *Biochemistry* 55, 373–381. doi:10.1021/acs.biochem.5b01207

Aravind, L., Wolf, Y. I., and Koonin, E. V. (2000). The ATP-Cone: an Evolutionarily mobile, ATP-Binding Regulatory Domain. *J. Mol. Microbiol. Biotechnol.* 2, 191–194. Available at: http://www.ncbi.nlm.nih.gov/entrez/query.fcgi?cmd=Retrieve&db=PubMed&dopt=Citation&list_uids=10939243.

Brignole, E. J., Tsai, K.-L., Chittuluru, J., Li, H., Aye, Y., Penczek, P. A., *et al.* (2018). 3.3-Å Resolution Cryo-EM Structure of Human Ribonucleotide Reductase with Substrate and Allosteric Regulators Bound. *Elife* 7, e31502. Available at: http://eutils.ncbi.nlm.nih.gov/entrez/eutils/elink.fcgi?dbfrom=pubmed&id=29460780&retmode=ref&cmd=prlinks

Chen, V. B., Arendall, W. B., 3rd, Headd, J. J., Keedy, D. A., Immormino, R. M., Kapral, G. J., *et al.* (2010). MolProbity: All-Atom Structure Validation for Macromolecular Crystallography. *Acta Crystallogr. D Biol. Cryst.* 66, 12–21. doi:10.1107/S0907444909042073

Fairman, J. W., Wijerathna, S. R., Ahmad, M. F., Xu, H., Nakano, R., Jha, S., *et al.* (2011). Structural Basis for Allosteric Regulation of Human Ribonucleotide Reductase by Nucleotide-Induced Oligomerization. *Nat. Struct. Mol. Biol.* 18, 316–322. doi:10.1038/nsmb.2007

Franke, D., and Svergun, D. I. (2009). DAMMIF, a Program for Rapidab-Initioshape Determination in Small-Angle Scattering. *J. Appl. Cryst.* 42, 342–346. Available at: http://scripts.iucr.org/cgi-bin/paper?S0021889809000338

Gasteiger, E., Hoogland, C., Gattiker, A., Duvaud, S. e., Wilkins, M. R., Appel, R. D., *et al.* (2005). "Protein Identification and Analysis Tools on the ExPASy Server," in *The Proteomics Protocols Handbook*, Totowa, NJ: Humana Press. 571–607. doi:10.1385/1-59259-890-0:571

Grant, T. D. (2018). Ab Initio electron Density Determination Directly from Solution Scattering Data. *Nat. Methods* 15, 191–193. doi:10.1038/nmeth.4581

Greene, B. L., Kang, G., Cui, C., Bennati, M., Nocera, D. G., Drennan, C. L., *et al.* (2020). Ribonucleotide Reductases: Structure, Chemistry, and Metabolism Suggest New Therapeutic Targets. *Annu. Rev. Biochem.* 89, 45–75. doi:10.1146/annurev-biochem-013118-111843

Guinier, A. (1939). La diffraction des rayons X aux très petits angles: application à l'étude de phénomènes ultramicroscopiques. *Ann. Phys.* 11, 161–237. doi:10.1051/anphys/193911120161

Hofer, A., Crona, M., Logan, D. T., and Sjöberg, B.-M. (2012). DNA Building Blocks: Keeping Control of Manufacture. *Crit. Rev. Biochem. Mol. Biol.* 47, 50–63. doi:10.3109/10409238.2011.630372

Högbom, M., Sjöberg, B. M., and Berggren, G. (2020). Radical Enzymes. *eLS* 1, 375–393. doi:10.1002/9780470015902.a0029205

Johansson, R., Jonna, V. R., Kumar, R., Nayeri, N., Lundin, D., Sjöberg, B.-M., *et al.* (2016). Structural Mechanism of Allosteric Activity Regulation in a Ribonucleotide Reductase with Double ATP Cones. *Structure* 24, 906–917. doi:10.1016/j.str.2016.03.025

Jonna, V. R., Crona, M., Rofougaran, R., Lundin, D., Johansson, S., Brännström, K., *et al.* (2015). Diversity in Overall Activity Regulation of Ribonucleotide Reductase. *J. Biol. Chem.* 290, 17339–17348. doi:10.1074/jbc.M115.649624

Kang, G., Taguchi, A. T., Stubbe, J., and Drennan, C. L. (2020). Structure of a Trapped Radical Transfer Pathway within a Ribonucleotide Reductase Holocomplex. *Science* 368, 424–427. doi:10.1126/science.aba6794

Kelley, L. A., Mezulis, S., Yates, C. M., Wass, M. N., and Sternberg, M. J. E. (2015). The Phyre2 Web portal for Protein Modeling, Prediction and Analysis. *Nat. Protoc.* 10, 845–858. doi:10.1038/nprot.2015.053

Konarev, P. V., Volkov, V. V., Sokolova, A. V., Koch, M. H. J., and Svergun, D. I. (2003). PRIMUS: a Windows PC-Based System for Small-Angle Scattering Data Analysis. *J. Appl. Cryst.* 36, 1277–1282. doi:10.1107/S0021889803012779

Lundin, D., Berggren, G., Logan, D., and Sjöberg, B.-M. (2015). The Origin and Evolution of Ribonucleotide Reduction. *Life* 5, 604–636. doi:10.3390/life5010604

Mathews, C. K. (2018). Still the Most Interesting Enzyme in the World. *FASEB j.* 32, 4067–4069. Available at: https://www.fasebj.org/doi/10.1096/fj.201800790R

Panjkovich, A., and Svergun, D. I. (2018). CHROMIXS: Automatic and Interactive Analysis of Chromatography-Coupled Small-Angle X-ray Scattering Data. *Bioinformatics* 34, 1944–1946. doi:10.1093/bioinformatics/btx846

Parker, M. J., Maggiolo, A. O., Thomas, W. C., Kim, A., Meisburger, S. P., Ando, N., *et al.* (2018). An Endogenous dAMP Ligand in Bacillus Subtilis Class Ib RNR Promotes Assembly of a Noncanonical Dimer for Regulation by dATP. *Proc. Natl. Acad. Sci. USA* 115, E4594–E4603. Available at: https://www.pnas.org/content/115/20/E4594.abstract

Pernot, P., Round, A., Barrett, R., De Maria Antolinos, A., Gobbo, A., Gordon, E., *et al.* (2013). Upgraded ESRF BM29 Beamline for SAXS on Macromolecules in Solution. *J. Synchrotron Radiat.* 20, 660–664. doi:10.1107/S0909049513010431

Petoukhov, M. V., Franke, D., Shkumatov, A. V., Tria, G., Kikhney, A. G., Gajda, M., *et al.* (2012). New Developments in the ATSAS program Package for Small-Angle Scattering Data Analysis. *J. Appl. Cryst.* 45, 342–350. doi:10.1107/S0021889812007662

Pettersen, E. F., Goddard, T. D., Huang, C. C., Couch, G. S., Greenblatt, D. M., Meng, E. C., *et al.* (2004). UCSF Chimera? A Visualization System for Exploratory Research and Analysis. *J. Comput. Chem.* 25, 1605–1612. doi:10.1002/jcc.20084

Punjani, A. (2020). Algorithmic Advances in Single Particle Cryo-EM Data Processing Using CryoSPARC. *Microsc. Microanal.* 26, 2322–2323. doi:10.1017/s1431927620021194

Punjani, A., Rubinstein, J. L., Fleet, D. J., and Brubaker, M. A. (2017). cryoSPARC: Algorithms for Rapid Unsupervised Cryo-EM Structure Determination. *Nat. Methods* 14, 290–296. Available at: http://www.nature.com/articles/nmeth.4169

Rofougaran, R., Vodnala, M., and Hofer, A. (2006). Enzymatically Active Mammalian Ribonucleotide Reductase Exists Primarily as an $\alpha_6\beta_2$ Octamer. *J. Biol. Chem.* 281, 27705–27711. Available at: http://www.ncbi.nlm.nih.gov/entrez/query.fcgi?cmd=Retrieve&db=PubMed&dopt=Citation&list_uids=16861739

Rose, H. R., Maggiolo, A. O., McBride, M. J., Palowitch, G. M., Pandelia, M.-E., Davis, K. M., *et al.* (2019). Structures of Class Id Ribonucleotide Reductase Catalytic Subunits Reveal a Minimal Architecture for Deoxynucleotide Biosynthesis. *Biochemistry* 58, 1845–1860. doi:10.1021/acs.biochem.8b01252

Rozman Grinberg, I., Berglund, S., Hasan, M., Lundin, D., Ho, F. M., Magnuson, A., *et al.* (2019). Class Id Ribonucleotide Reductase Utilizes a Mn_2(IV,III) Cofactor and Undergoes Large Conformational Changes on Metal Loading. *J. Biol. Inorg. Chem.* 24, 863–877. doi:10.1007/s00775-019-01697-8

Rozman Grinberg, I., Lundin, D., Hasan, M., Crona, M., Jonna, V. R., Loderer, C., *et al.* (2018a). Novel ATP-Cone-Driven Allosteric Regulation of Ribonucleotide Reductase via the Radical-Generating Subunit. *Elife* 7, 1–26. doi:10.7554/eLife.31529

Rozman Grinberg, I., Lundin, D., Sahlin, M., Crona, M., Berggren, G., Hofer, A., *et al.* (2018b). A Glutaredoxin Domain Fused to the Radical-Generating Subunit of Ribonucleotide Reductase (RNR) Functions as an Efficient RNR Reductant. *J. Biol. Chem.* 293, 15889–15900. doi:10.1074/jbc.ra118.004991 Available at: http://www.jbc.org/lookup/doi/10.1074/jbc.RA118.004991

Shukla, A. K., Westfield, G. H., Xiao, K., Reis, R. I., Huang, L.-Y., Tripathi-Shukla, P., *et al.* (2014). Visualization of Arrestin Recruitment by a G-Protein-Coupled Receptor. *Nature* 512, 218–222. doi:10.1038/nature 13430

Svergun, D. I., Petoukhov, M. V., and Koch, M. H. J. (2001). Determination of Domain Structure of Proteins from X-ray Solution Scattering. *Biophysical J.* 80, 2946–2953. doi:10.1016/s0006-3495(01)76260-1 Available at: http://www.ncbi.nlm.nih.gov/pmc/articles/PMC1301478/pdf/113 71467.pdf

Thomas, W. C., Brooks, F. P., Burnim, A. A., Bacik, J.-P., Stubbe, J., Kaelber, J. T., *et al.* (2019). Convergent Allostery in Ribonucleotide Reductase. *Nat. Commun.* 10, 1–13. doi:10.1038/s41467-019-10568-4

Torrents, E. (2014). Ribonucleotide Reductases: Essential Enzymes for Bacterial Life. *Front. Cel. Infect. Microbiol.* 4, 52. doi:10.3389/fcimb. 2014.00052

Uhlin, U., and Eklund, H. (1994). Structure of Ribonucleotide Reductase Protein R1. *Nature* 370, 533–539. doi:10.1038/370533a0

Volkov, V. V., and Svergun, D. I. (2003). Uniqueness of *ab initio* shape Determination in Small-Angle Scattering. *J. Appl. Cryst.* 36, 860–864. doi:10.1107/S0021889803000268

Waterhouse, A., Bertoni, M., Bienert, S., Studer, G., Tauriello, G., Gumienny, R., *et al.* (2018). SWISS-MODEL: Homology Modelling of Protein Structures and Complexes. *Nucleic Acids Res.* 46, W296–W303. doi:10.1093/nar/gky427

Wriggers, W. (2012). Conventions and Workflows for Using Situs. *Acta Crystallogr. D Biol. Cryst.* 68, 344–351. doi:10.1107/S0907444911049791

Yang, J., Yan, R., Roy, A., Xu, D., Poisson, J., and Zhang, Y. (2015). The I-TASSER Suite: Protein Structure and Function Prediction. *Nat. Methods* 12, 7–8. doi:10.1038/nmeth.3213

Conflict of Interest: The authors declare that the research was conducted in the absence of any commercial or financial relationships that could be construed as a potential conflict of interest.

Publisher's Note: All claims expressed in this article are solely those of the authors and do not necessarily represent those of their affiliated organizations, or those of the publisher, the editors and the reviewers. Any product that may be evaluated in this article, or claim that may be made by its manufacturer, is not guaranteed or endorsed by the publisher.

Chapter 7

Nucleotide Binding to the ATP-cone in Anaerobic Ribonucleotide Reductases Allosterically Regulates Activity by Modulating Substrate Binding*

Ornella Bimai[1], Ipsita Banerjee [2], Inna Rozman Grinberg[1], Ping Huang[3], Lucas Hultgren[4], Simon Ekström[4], Daniel Lundin[1], Britt-Marie Sjöberg[1], Derek T Logan[2,5]

[1]*Department of Biochemistry and Biophysics, Stockholm University, Stockholm, Sweden*
[2]*Section for Biochemistry and Structural Biology, Centre for Molecular Protein Science, Department of Chemistry, Lund University, Lund, Sweden*
[3]*Department of Chemistry — Ångström Laboratory, Uppsala University, Uppsala, Sweden*
[4]*Structural Proteomics, SciLifeLab, Lund University, Lund, Sweden*
[5]*Cryo-EM for Life Science, SciLifeLab, Lund University, Lund, Sweden*

A small, nucleotide-binding domain, the ATP-cone, is found at the N-terminus of most ribonucleotide reductase (RNR) catalytic subunits. By binding adenosine triphosphate (ATP) or deoxyadenosine triphosphate (dATP) it regulates the enzyme activity of all classes of RNR. Functional and structural work on aerobic RNRs has revealed a plethora

*The chapter was originally published in Ornella Bimai, Ipsita Banerjee, Inna Rozman Grinberg, Ping Huang, Lucas Hultgren, Simon Ekström, Daniel Lundin, Britt-Marie Sjöberg, Derek T Logan (2023) Nucleotide Binding to the ATP-cone in Anaerobic Ribonucleotide Reductases Allosterically Regulates Activity by Modulating Substrate Binding. *eLife* 12:RP89292. doi: https://doi.org/10.7554/eLife.89292.

of ways in which dATP inhibits activity by inducing oligomerisation and preventing a productive radical transfer from one subunit to the active site in the other. Anaerobic RNRs, on the other hand, store a stable glycyl radical next to the active site and the basis for their dATP-dependent inhibition is completely unknown. We present biochemical, biophysical, and structural information on the effects of ATP and dATP binding to the anaerobic RNR from *Prevotella copri*. The enzyme exists in a dimertetramer equilibrium biased towards dimers when two ATP molecules are bound to the ATP-cone and tetramers when two dATP molecules are bound. In the presence of ATP, *P. copri* NrdD is active and has a fully ordered glycyl radical domain (GRD) in one monomer of the dimer. Binding of dATP to the ATP-cone results in loss of activity and increased dynamics of the GRD, such that it cannot be detected in the cryo-EM structures. The glycyl radical is formed even in the dATP-bound form, but the substrate does not bind. The structures implicate a complex network of interactions in activity regulation that involve the GRD more than 30 Å away from the dATP molecules, the allosteric substrate specificity site and a conserved but previously unseen flap over the active site. Taken together, the results suggest that dATP inhibition in anaerobic RNRs acts by increasing the flexibility of the flap and GRD, thereby preventing both substrate binding and radical mobilisation.

eLife assessment

This study advances our understanding of the allosteric regulation of anaerobic ribonucleotide reductases (RNRs) by nucleotides, providing **valuable** new structural insight into class III RNRs containing ATP cones. The cryo-EM structural characterization of the system is **solid**, but some open questions remain about the interpretation of activity/binding assays and the HDX-MS results that have been newly incorporated compared to a previous version. The work will be of interest to biochemists and structural biologists working on ribonucleotide reductases and other allosterically regulated enzymes.

Introduction

Ribonucleotide reductases (RNRs) are a family of enzymes with sophisticated radical chemistries and allosteric regulation. RNRs produce all four deoxyribonucleotides and are the only enzymes providing de novo building blocks for DNA replication and repair in all free-living organisms (Mathews, 2016). Virtually all RNRs possess an allosteric site regulating substrate specificity (the s-site), a crucial aspect of RNR that ultimately provides a balanced supply of deoxyribonucleoside triphosphates (dNTPs) (Mathews, 2016; Mathews, 2018). A second allosteric site (the a-site) is located in an N-terminal domain called the ATP-cone (Aravind *et al.*, 2000), found in the majority of RNRs (Jonna *et al.*, 2015). Whereas the s-site binds all dNTPs and usually ATP, the a-site in the ATP-cone can only bind ATP and dATP. Binding of ATP activates RNR, and binding of dATP inhibits it (Hofer *et al.*, 2012; Martínez-Carranza *et al.*, 2020).

The RNR family consists of three evolutionarily related classes (I-III) with a common radical-based reaction mechanism but differing in the mode of radical generation and in quaternary structure (Wiley, 2001; Lundin *et al.*, 2015). Class I, strictly aerobic RNRs are heterotetramers consisting of two α subunits (NrdA) with binding sites for substrates and allosteric nucleotides and two β subunits (NrdB) that harbour a radical initiator in the form of a stable radical or an oxidised metal cluster (Wiley, 2001). Class II RNRs are either dimers or monomers with an α subunit (NrdJ) that, in addition to binding sites for substrate and allosteric nucleotides, also harbours the radical initiator cofactor adenosylcobalamin. Class III RNRs, which only function anaerobically, are dimers of an α subunit (NrdD) with a stable radical close to the active site, and binding sites for allosteric nucleotides. The radical is introduced via encounter with a specific radical-SAM enzyme called NrdG, and once activated NrdD can perform multiple turnovers in the absence of NrdG (Backman *et al.*, 2017; Torrents *et al.*, 2001).

The dATP-dependent inhibition has been biochemically characterised in representatives of all three RNR classes, but only structurally studied in class I. A common denominator of all inhibited class I RNRs is oligomerisation, leading to disturbed radical transfer between the α and β subunits. Hitherto four different inhibitory oligomerisation mechanisms have been identified: heterooctameric $\alpha_4\beta_4$ complexes in *Escherichia coli, Clostridium botulinum*, and *Neisseria gonorrhoeae* (Ando *et al.*, 2011; Martínez-Carranza *et al.*, 2020;

Wei *et al.*, 2014), an α_4 complex in *Pseudomonas aeruginosa* (Johansson *et al.*, 2016), α_6 complexes in the eukaryotes *Homo sapiens, Saccharomyces cerevisiae*, and *Dictyostelium discoideum* (Ando *et al.*, 2016; Brignole *et al.*, 2018; Crona *et al.*, 2013; Fairman *et al.*, 2011), and a β_4 complex in the bacterium *Leeuwenhoekiella blandensis* (Rozman Grinberg *et al.*, 2018a). All of these class I oligomers involve protein-protein interactions mediated by the ATP-cone domain.

The ATP-cone is in essence restricted to RNRs and the RNR-specific repressor NrdR (Rozman Grinberg *et al.*, 2022). Among RNRs it is found in about 80% of class III enzymes and about 50% of class I enzymes, but less than 10% of class II enzymes (Jonna *et al.*, 2015). The presence and function of the ATP-cone domain distinguish anaerobic RNRs from the other members of the large glycyl radical enzyme (GRE) family that are otherwise structurally related (Backman *et al.*, 2017). The allosteric regulation mechanisms of class III RNRs (NrdDs) from *E. coli*, bacteriophage T4 and *Lactococcus lactis* were determined several decades ago (Andersson *et al.*, 2000; Eliasson *et al.*, 1994; Torrents *et al.*, 2000). T4NrdD was the first class III RNR structure solved (Logan *et al.*, 1999), and later, structures of *Thermotoga maritima* NrdD (TmNrdD) have been published (Aurelius *et al.*, 2015; Wei *et al.*, 2014). However, these two NrdDs lack an ATP-cone and the structural basis for allosteric activity regulation in the class III RNRs is still an outstanding question. Here, we present structural, biochemical, and biophysical studies on the ATP-cone-containing NrdD from the human pathogen *Prevotella copri* (PcNrdD; Nii *et al.*, 2023), showing that the binding of dATP causes increased flexibility of the C-terminal glycyl radical-bearing domain, as well as a flap-like loop that binds across the top of the active site in the active form. This increase in dynamics is coupled to enzymatic inactivation by inhibition of substrate binding and radical transfer. The final outcome of dATP inhibition is thus blocked radical transfer in both class I and III RNRs, but it is achieved by completely different mechanisms, i.e. oligomerisation prevents radical transfer between subunits in class I RNRs but long-range modulation of the flexibility of key structural elements prevents substrate binding in class III RNRs.

Results

Allosteric activity regulation by ATP and dATP

The amino acid sequence of PcNrdD suggests that it belongs to the formate-requiring class III RNRs (Burnim *et al.*, 2022b; Wei *et al.*, 2014), and initial optimisation of the assay composition showed that this was the case (Figure 1 — figure supplement 1). ATP-cone-mediated activation of PcNrdD enzyme activity by ATP or inhibition by dATP was tested with two different substrates. With guanosine triphosphate (GTP) as substrate the s-site was filled with dTTP as effector, and in absence of any a-site effector the basal level of GTP reduction was 21 ± 2 nmol/min/mg. An increasing concentration of ATP stimulated activity to a k_{cat} of $1.3\,s^{-1}$(478 ± 26 nmol/min/mg), with a K_L of 0.67 ± 0.12 mM (Figure 1a). On the other hand, an increasing concentration of dATP resulted in an abrupt inhibition of enzyme activity with a K_i of $74 \pm 24\,\mu$M (Figure 1b). Hence, the K_i value for dATP inhibition via the ATP-cone is approximately 10-fold lower compared to the K_L value for ATP activation via the ATP-cone.

The activation/inhibition experiments were also carried out with CTP as substrate. Specificity effectors for CTP reduction are ATP or dATP, so in these experiments both the s- and a-sites were conceivably filled with ATP or dATP, respectively. It is obvious from Figure 1c that increasing concentrations of ATP initially stimulate activity and that high levels of ATP cause inhibition, plausibly by ATP also acting as substrate. We therefore used only results up to 3 mM ATP to calculate an approximate k_{cat} of $1.1\,s^{-1}$(385 ± 116 nmol/min/mg) with an apparent K_L of 0.67 ± 0.52 mM (Figure 1c). Addition of dATP initially results in a stimulation of enzyme activity, conceivably when the s-site is filled with dATP, after which competing inhibition appears when dATP also binds to the ATP-cone (Figure 1d). All in all, these results suggest that the ATP-cone

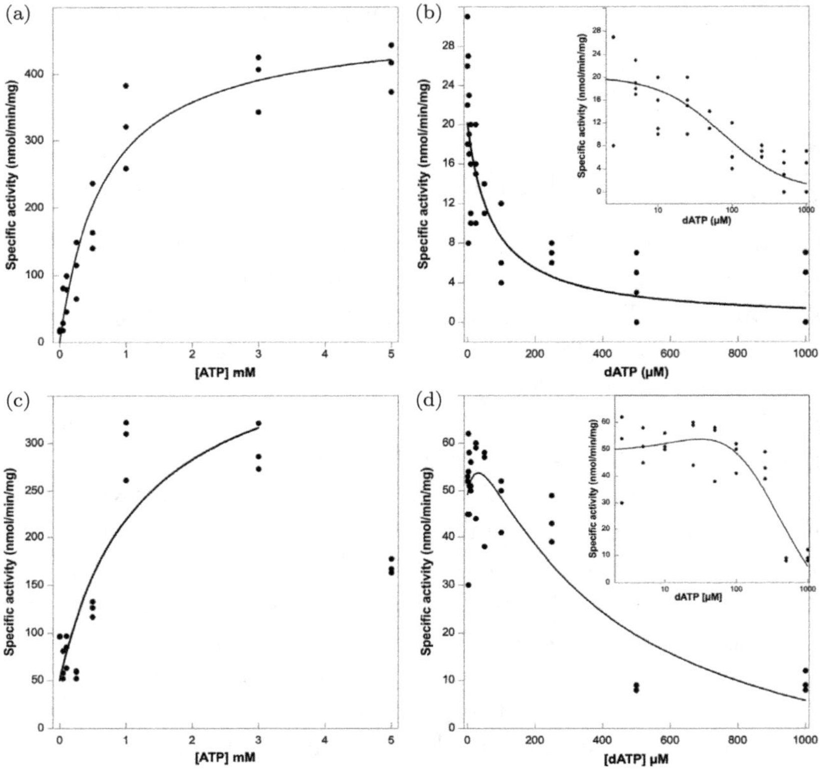

Figure 1: Activity assays of *Prevotella copri* NrdD in the presence of ATP or dATP. A 2.5-fold excess of holoNrdG over apoNrdD was used to study the allosteric regulation effect of ATP and dATP on the a-site. GTP reduction was monitored with 1 mM dTTP as effector in the s-site and titrated with ATP (a) or dATP (b) in the a-site. Cytidine triphosphate (CTP) reduction assays were titrated with ATP (c) or dATP (d), in this case acting both as s-site effectors and a-site regulators. Experiments were performed in triplicate for (a), (c), and (d) and with four replicates for (b). Insets in panels (b) and (d) show the results plotted in log scale. Curve fits for calculation of K_L and K_i used Equations 1 and 2, respectively, given in Materials and methods, and in (d) Equation 3. Curve fits for (c) and (d) used a start activity of 50 nmol/min/mg, and in (c) only results for 0–3 mM ATP were used. The R values for curve fits in panels (a–d) were 0.99, 0.95, 0.83, and 0.96, respectively.
The online version of this article includes the following figure supplement(s) for figure 1:
Figure supplement 1. Formate requirement in PcNrdD.
Figure supplement 2. Reconstitution of the [Fe–S] cluster of the *Prevotella copri* NrdG.

in PcNrdD responds similarly to inhibition by dATP as has been observed before for several class I RNRs.

Binding of nucleotides to the ATP-cone

To confirm binding of ATP or dATP to the ATP-cone we used isothermal titration calorimetry (ITC) and microscale thermophoresis (MST) (Figure 2). Substrate GTP and s-site effector dTTP were used to fill up the other nucleotide-binding sites in PcNrdD. Binding of dATP occurred with a K_D of 6 μM, whereas binding of ATP was fourfold weaker with a K_D of 26 μM (Figure 2a–c). The number of bound dATP molecules observed with ITC was almost twofold higher for dATP compared to ATP, but both values were below 1. Nucleotide binding measured with MST confirmed our results for ATP (Figure 2d) that had a K_D of 25 μM, and dATP that with this method had a K_D of 23 μM (Figure 2e). However, in cryo-EM experiments, where higher nucleotide concentrations can be used, we show that the ATP-cone can bind two ATP or two dATP molecules (see below), as also expected based on its sequence. The K_L and K_i values reported above for ATP activation and dATP inhibition of GTP reduction conceivably reflect the requirement for two bound nucleotide molecules to achieve these effects, whereas the K_D measurements via ITC and MST are restricted by the component concentrations that can be used and plausibly only reflect the first nucleotide bound.

It is noteworthy that the K_D values for ATP and dATP do not differ by orders of magnitude, as has been found for several class I RNRs (Ando *et al.*, 2016; Birgander *et al.*, 2004; Ormö and Sjöberg, 1990; Rozman Grinberg *et al.*, 2018a; Rozman Grinberg *et al.*, 2018b; Torrents *et al.*, 2006). Collectively, ITC and MST instead suggested that ATP and dATP may interact at two distinct sites in the ATP-cone for the functional readout of dATP inhibition, as has earlier been observed in the transcriptional repressor NrdR (Rozman Grinberg *et al.*, 2022). To test this hypothesis, apo-PcNrdD was incubated with ATP only, dATP only, or a combination of ATP and dATP, then desalted, boiled, centrifuged, and the nucleotide content of the

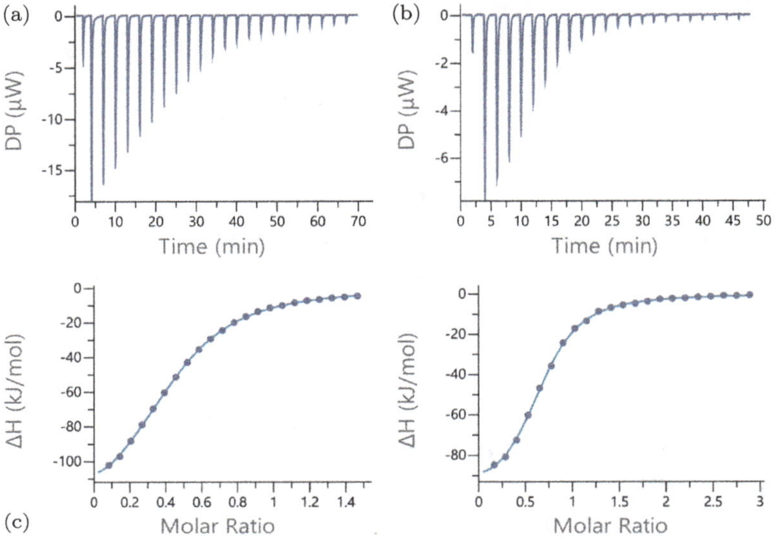

(c)

Ligand	N (sites)	K_D (μM)	ΔH (kJ/mol)	ΔG (kJ/mol)	-TΔS (kJ/mol)
ATP	0.42±0.04	26±6	-150±17	-26±0.5	125±17
dATP	0.74±0.10	5.8±3	-118±18	-30±1.1	88±19

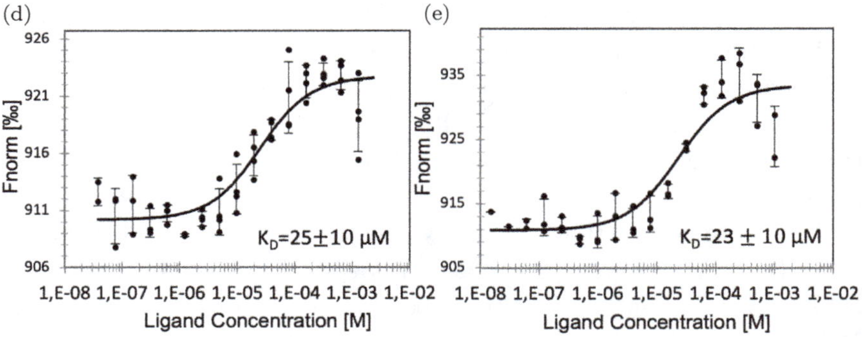

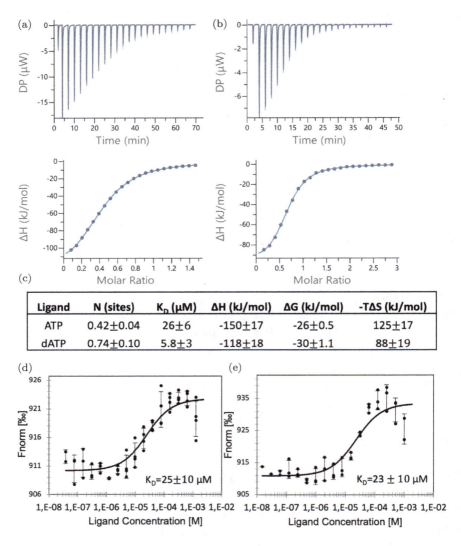

Figure 2: Binding of nucleotides to the ATP-cone. Binding assay using isothermal titration calorimetry (ITC): Representative thermograms of binding of ATP to 50 μM PcNrdD (a); and binding of dATP to 200 μM PcNrdD (b); (c) thermodynamic parameters of ligand binding. All titrations were performed in the presence of 1 mM GTP substrate and 1 mM s-site effector dTTP at 20°C. Binding of nucleotides to the ATP-cone using microscale thermophoresis (MST): (d) Binding of ATP and (e) binding of dATP. All MST-binding experiments were performed in the presence of 1 mM s-site effector dTTP and 5 mM GTP substrate at room temperature. Control experiments were done for ITC, by titrating the nucleotide into the buffer (not shown).

supernatant analysed by high performance liquid chromatography (HPLC). Table 1 shows that only dATP was bound when PcNrdD was incubated with a combination of ATP and dATP. This was also the case when incubations were performed in the presence of s-site effector and substrate, and also when the desalting was performed in the presence of s-site allosteric effector and substrate. In incubations with only ATP, s-site effector and substrate were needed during the entire work-up procedure for ATP to be retained by the protein. In conclusion, the binding experiments suggest that the PcNrdD ATP-cone can bind either ATP or dATP but not both simultaneously. Importantly, the results point to allosteric communication between the s-site, the active site, and the a-site in the ATP-cone, as dTTP and GTP are required for any ATP to be retained. Both of these observations are consistent with the structural analyses shown below.

Modulation of the oligomeric state by ATP and dATP in the presence of allosteric effectors

Binding of dATP to the ATP-cone in all class I RNRs studied to date results in formation of higher oligomers unable to perform catalysis (reviewed in Martínez-Carranza *et al.*, 2020). We therefore next asked whether a similar mechanism would be valid for dATP-mediated inhibition of a class III RNR. To elucidate the oligomeric states of PcNrdD we used gas-phase electrophoretic mobility molecular analysis (GEMMA) (Kaufman *et al.*, 1996), a method similar to electrospray mass spectrometry. It requires low protein and nucleotide concentrations as well as volatile buffers.

When loaded with ATP, PcNrdD was in a dimer-tetramer equilibrium shifted towards dimers (Figure 3a; Figure 3 — figure supplement 1; Table 2). Addition of the CTP substrate or the s-site effector dTTP resulted in a similar equilibrium. In contrast, the dimer-tetramer equilibrium of dATP-loaded PcNrdD was shifted towards tetramers, and this equilibrium was likewise not affected by addition of the CTP substrate (Figure 3b; Figure 3 — figure supplement 1, Table 2). Hence, dATP inhibition of enzyme activity seems to work differently in PcNrdD than the clear-cut and drastic

Table 1: Amount of ATP and/or dATP bound to PcNrdD under different incubation and desalting condition.

Incubating conditions	**Desalting conditions**	**ATP (mol/mol NrdD)**	**dATP (mol/mol NrdD)**
NrdD + ATP*		0.09***	0
NrdD + dATP†	No nucleotides during desalting	0	0.97†††
NrdD + ATP + dATP‡		0.01	0.89‡‡‡
NrdD + ATP + dTTP + GTP§		0.03	0.01
NrdD + dATP + dTTP + GTP¶		0	0.64
NrdD + ATP + dATP + dTTP + GTP**		0.01	1
NrdD + ATP + dTTP + GTP††	No nucleotides during desalting	0.35	0
NrdD + dATP + dTTP + GTP‡‡	dTTP + GTP included during desalting¶¶	0	0.52
NrdD + ATP + dATP + dTTP + GTP§§		0.05	0.85

*1 mM ATP.
†1 mM dATP.
‡1 mM ATP, 1 mM dATP.
§3 mM ATP, 1 mM dTTP, 5 mM GTP.
¶1 mM dATP, 1 mM dTTP, 5 mM GTP
**3 mM ATP, 1 mM dATP, 1 mM dTTP, 5 mM GTP.
††3 mM ATP, 2 mM dTTP, 5 mM GTP.
‡‡1 mM dATP, 2 mM dTTP, 5 mM GTP.
§§3 mM ATP, 1 mM dATP, 2 mM dTTP, 5 mM GTP.
¶¶0.1 mM dTTP, 1 mM GTP.
***Summary of 0.07 mol/mol NrdD ATP and 0.02 mol/mol NrdD ADP.
†††Summary of 0.78 mol/mol NrdD dATP, 0.18 mol/mol NrdD dADP, and 0.01 mol/mol NrdD dAMP.
‡‡‡Summary of 0.70 mol/mol NrdD dATP, 0.18 mol/mol NrdD dADP, and 0.01 mol/mol NrdD dAMP.

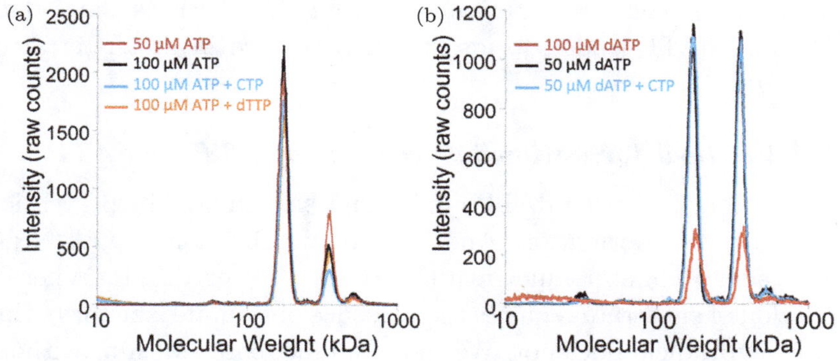

Figure 3: Oligomeric states of *Prevotella copri* NrdD in the presence of nucleotides determined by gas-phase electrophoretic mobility molecular analysis (GEMMA). (a) Apo-PcNrdD ($2\,\mu$M) loaded with the activator ATP (50–100 μM) (black and red) in the presence of CTP ($100\,\mu$M) as a substrate (cyan) or dTTP ($100\,\mu$M; orange) as the allosteric effector. (b) apo-PcNrdD ($2\,\mu$M) loaded with the inhibitor dATP (50–100 μM) (black and red) in the presence of CTP ($50\,\mu$M) (cyan) as the substrate. Each sample was scanned five times to increase signal-to-noise level. (c) Calculated numbers of monomers based on measured molecular weight and fractions of dimers versus tetramers after conversion of experiments (a) and (b) to mass concentrations.
The online version of this article includes the following figure supplement(s) for figure 3:
Figure supplement 1. Size exclusion chromatography (SEC) analyses of *Prevotella copri* NrdD in the presence of nucleotides.

Table 2: Calculated numbers of monomers based on measured molecular weight and fractions of dimers versus tetramers after conversion of experiments (a) and (b) to mass concentrations.

Effector(s)	**Mw (kDa)**	**No. of monomers**	**Dimers vs. tetramers (%)**
ATP ($50\,\mu$M)	176/358	2.1/4.2	50/50
ATP ($100\,\mu$M)	178/360	2.1/4.3	60/40
ATP ($100\,\mu$M) + CTP	179/356	2.1/4.3	64/36
ATP ($100\,\mu$M) + dTTP	171/348	2.1/4.2	60/40
dATP ($50\,\mu$M)	167/363	2.0/4.3	32/68
dATP ($100\,\mu$M)	179/361	2.1/4.3	32/68
dATP ($50\,\mu$M) + CTP	175/358	2.1/4.2	34/66

change in oligomeric structure observed for class I RNRs, since both ATP- and dATP-loaded forms of PcNrdD were in dimer-tetramer equilibria.

Glycyl radical formation in presence of ATP or dATP

Next, we asked whether dATP inhibition was mediated by prevention of glycyl radical formation. Apo-PcNrdD was incubated with NrdG and *S*-adenosylmethionine in presence of ATP or dATP. Whereas reduction of substrate requires the presence of formate, the formation of a glycyl radical does not. We therefore incubated with or without formate as well as with or without substrate. Figure 4 shows that the glycyl radical was readily formed in the presence of dATP both with and without the CTP substrate, and the result was similar in the absence of formate (Figure 4 — figure supplement 1). Interestingly, under all conditions the glycyl radical content was higher in the presence of dATP compared to with ATP, and it was also more stable than in the absence of any effector nucleotides. In addition, the shape of the doublet electron paramagnetic resonance (EPR) signal was similar under all conditions studied. Collectively, these results show that the glycyl radical is readily formed in the presence of a dATP concentration that inhibits RNR activity.

Substrate binding to PcNrdD

To study substrate binding to activated and inhibited PcNrdD we used MST with labelled protein and GTP or CTP as substrate. The results were clear-cut in both sets of experiments (Figure 5). ATP-loaded PcNrdD bound GTP in the presence of s-site effector dTTP (Figure 5a), and it bound CTP in the presence of only ATP (Figure 5b), whereas dATP-loaded PcNrdD had extremely low affinity to GTP and CTP under similar conditions (Figure 5c, d). These results show that the dATP inhibition of enzyme activity in a class III RNR is mediated by inhibited binding of substrate.

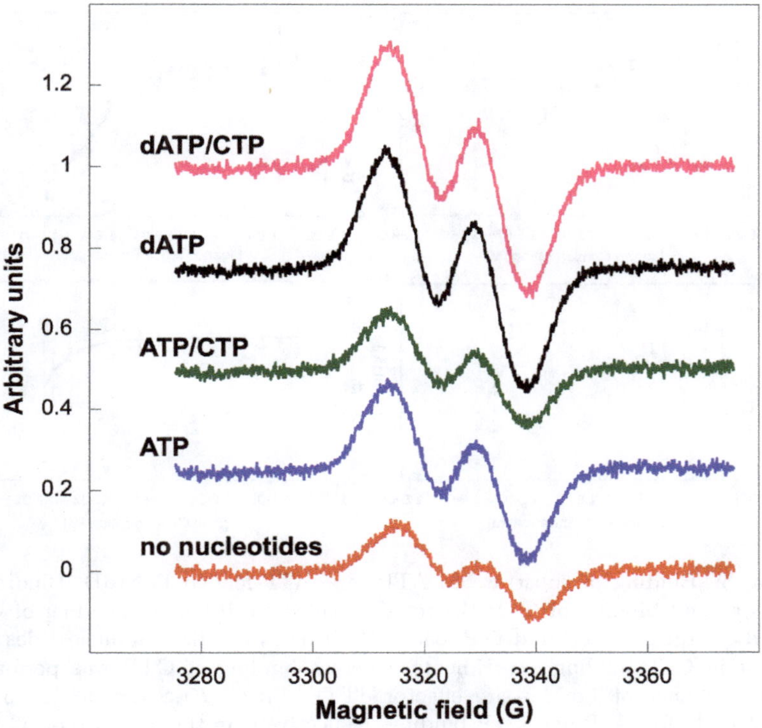

Figure 4: Glycyl radical formation after 20 min in presence of formate and ATP ± CTP or dATP ± CTP. Nucleotide concentrations were: 1.5 mM ATP, 1 mM dATP, and 1 mM CTP. Traces are arbitrarily moved to increase visibility and scaled to identical units (Y-axes).
The online version of this article includes the following figure supplement(s) for figure 4:
Figure supplement 1. Glycyl radical formation in absence of formate after 20-min incubation.

Cryo-EM structure of the ATP-CTP-bound PcNrdD dimer

Two-dimensional (2D) classes of particles extracted from micrographs of PcNrdD incubated with effector ATP and substrate CTP indicated only dimers, with no appreciable amounts of tetramer.

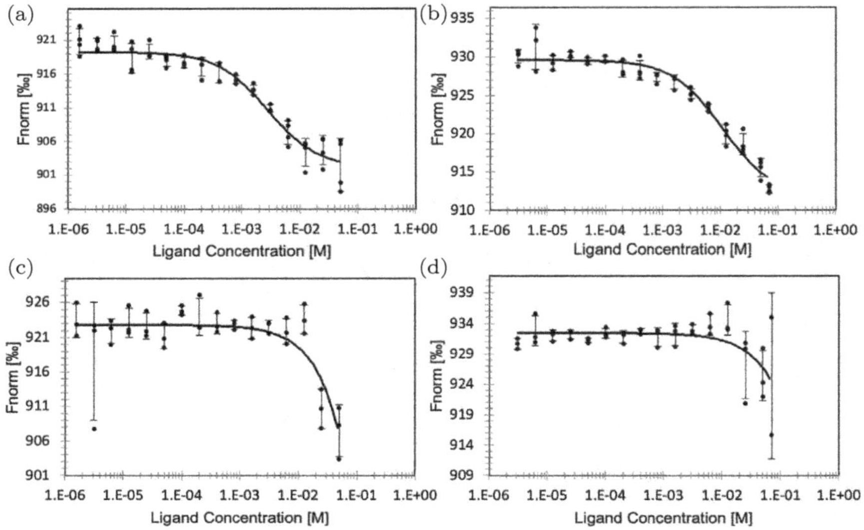

Figure 5: Binding of substrate to ATP- or dATP-loaded PcNrdD. Binding of GTP (a) and binding of CTP (b) to ATPloaded PcNrdD. (c) Binding of GTP and (d) binding of CTP to dATP-loaded PcNrdD. No additional nucleotides were present in CTP-binding experiments, whereas binding of GTP was performed in the presence of 1 mM s-site effector dTTP. Fitted K_DS are 2.8 ± 0.5 and 11 ± 2.4 mM for GTP and CTP binding, respectively, in the presence of ATP. In the presence of dATP fitted K_Ds were ≥ 2833 and ≥ 803 mM for GTP and CTP binding, respectively. Experiments were performed at room temperature.

From a set of 570, 730 particles we obtained a map from cryoSPARC at 3.0 Å. Application of deep learning local sharpening in DeepEMhancer (Sanchez-Garcia *et al.*, 2021) improved the quality of the map, and we were able to build a reliable model for the great majority of the structure. As the ATP-cone density still did not permit reliable modelling, three-dimensional (3D) classification was carried out on this particle set, enabling the identification of a subset of 291, 231 particles that gave a volume at 3.2 Å with better-defined ATP-cones (Figure 6a). This map was used for refinement of the final model (Figure 6 — figure supplement 1).

PcNrdD is a compact dimer (Figure 6b). Each monomer consists of 739 residues, of which residues 1–91 constitute the ATP-cone, 92–110 a linker region, 111–122 a flap that folds over the substrate in

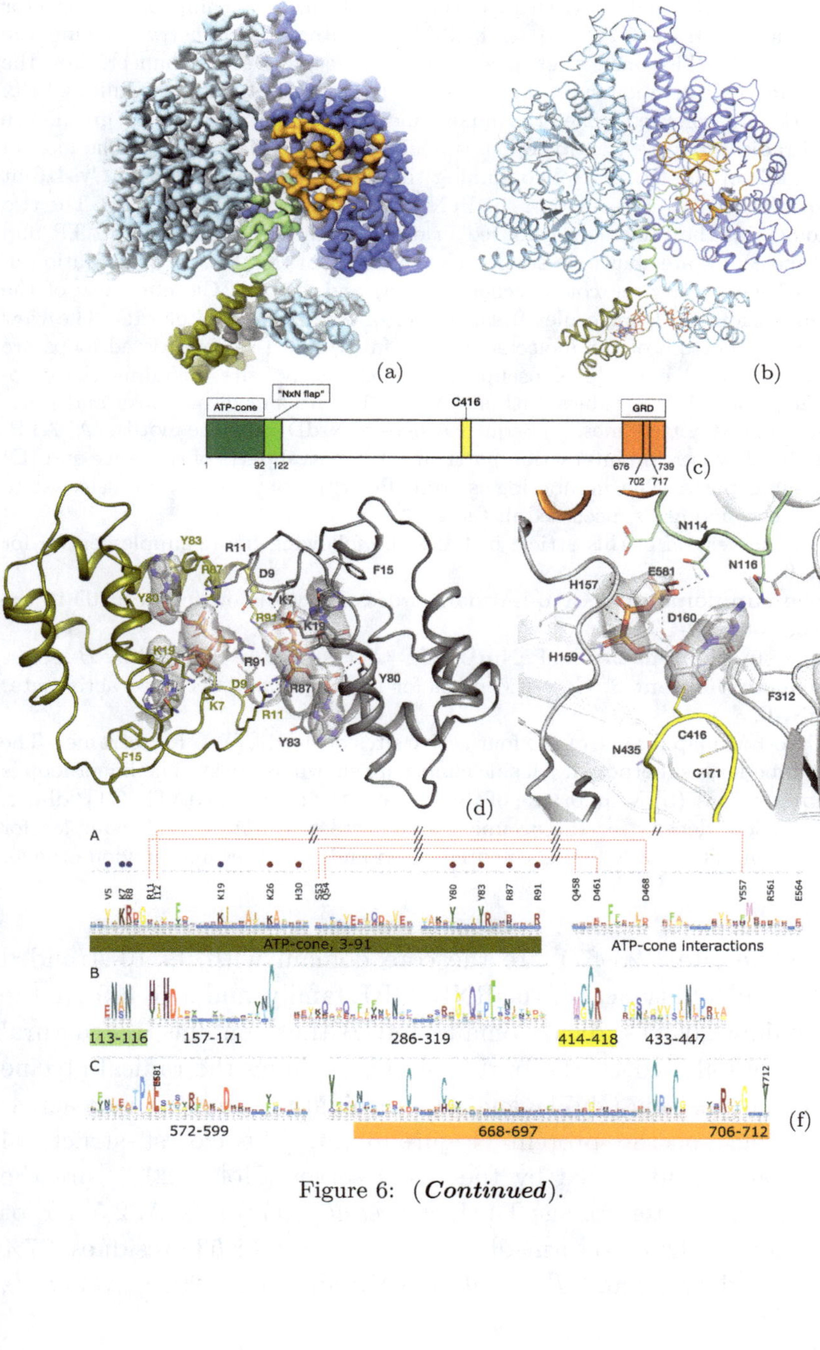

Figure 6: (***Continued***).

Figure 6 (***Continued***): Structure of the PcNrdD dimer in complex with effector ATP and substrate CTP. (a) Cryo-EM map with C1 symmetry showing the best-ordered ATP-cones after post-processing using DeepEMhancer. For the right-hand (active) monomer, the ATP-cone (residues 1–91) and the linking helix (92–104) are coloured olive, the linker and $N \times N$ flap region pale green, the glycyl radical loop red, and the C-terminal extended region orange. The loop in the middle of the α/β barrel containing the radical initiator cysteine Cys416 at its tip is yellow. (b) Overview of the PcNrdD dimer with ATP and CTP. The two monomers of the dimer are coloured in different shades of grey. The ATP and CTP molecules are shown as sticks. (c) Schematic of the domain organisation of PcNrdD with the same colour scheme as (a) and (b); (d) Closeup view of the binding of four ATP molecules to the dimer of ATP-cones in PcNrdD. The view is from the bottom of the molecule as seen in (a) and (b). Disordered loops are shown as dotted lines. (e) Closeup view of the active site including the cryo-EM map for CTP. Residues within 4 Å of CTP are shown as sticks and polar interactions as dotted lines. (f) Sequence logos of NrdD sequence motifs. (A) ATP-cone plus downstream interaction partners, (B) central parts of sequence and (C) C-terminal parts. The numbering is from PcNrdD. Segments were selected to illustrate amino acids discussed in the text.
The online version of this article includes the following figure supplement(s) for figure 6:
Figure supplement 1. Cryo-EM data processing workflow for PcNrdD in the presence of ATP-CTP.
Figure supplement 2. The PcNrdD-ATP-CTP structure coloured by *B*-factor.
Figure supplement 3. Cryo-EM maps for the empty and occupied active sites of PcNrdD.
(a) Cryo-EM map for one of the four active sites of the dATP-CTP tetramer. The refined model is superposed. All side chains are shown as sticks. The finger loop is coloured yellow. (b) Map for one of the two active sites of the dATP-CTP dimer. (c) Complete density for the occupied active site of the ATP-CTP complex for comparison. The $N \times N$ flap is coloured green and the C-terminal region orange.

the active site, 123–671 are the core domain with its 10-stranded α/β barrel fold typical of the RNR/GRE family, and 676–739 are the C-terminal glycyl radical domain (GRD) that contains a structural Zn^{2+} site followed by the buried loop containing the radical glycine residue at position Gly711 and a C-terminal tail that extends across the surface of the protein (Figure 6b,c). The closest structural neighbours as identified by the DALI server (Holm, 2022) are the NrdDs from bacteriophage T4 (Logan *et al.*, 1999) (1H7A, 2.5 Å root mean square deviation [rmsd] in C α positions for 535 residues, 27% sequence identity) and *T. maritima* (Aurelius *et al.*, 2015; Wei *et al.*,

Table 3: Structural similarity comparisons to PcNrdD. Structural similarity tables from DALI (PDB90, i.e. targets filtered at 90% sequence identity).

PDB ID	Protein	Z-score	Residues aligned	RMSD (Å)	Sequence identity (%)
1H7A	Bacteriophage T4 NrdD	34.6	535	2.5	27
4U3E	*T. maritima* NrdD	25.5	508	3.0	16
4COL	*T. maritima* NrdD/dATP	24.8	493	2.9	17

2014) (4U3E, 3.0 Å rmsd for 508 residues, 16% sequence identity; Table 3). The linker, flap, and GRD are well-ordered in one of the monomers (dark blue in Figure 6) and substrate CTP is bound (Figure 6e), while in the other monomer this region is less well-ordered and the active site is empty (Figure 6 — figure supplements 2 and 3). We will refer to these as the 'active' and 'inactive' monomers, respectively.

At the concentration of ATP used (1 mM), ATP is not observed to bind to the s-sites near the dimer interface. The ATP-cones containing the a-site are flexible, but the reconstruction from a subset of particles allowed the modelling of two ATP molecules per ATP-cone and many of the most important side chains. The ATP molecules are bound similarly to the twin dATP molecules seen in the ATP-cones of *P. aeruginosa* NrdA (PaNrdA) and *L. blandensis* NrdB (LbNrdB) (Johansson *et al.*, 2016; Rozman Grinberg *et al.*, 2018a), and to dATP bound to PcNrdD itself (see below). However, the ATP-cones in PcNrdD exhibit a hitherto unobserved relative orientation in which the four triphosphate groups are projected towards each other (Figure 6b,d). The large negative charge is compensated for by the proximity of as many as 12 Arg and Lys side chains from both monomers. The ATP-cones interact by contacts involving the 'roof' of the domain (residues 1–18). Arg11 of each ATP-cone reaches over to interact with the α-phosphate group of one of the ATP molecules in the other ATP-cone, and a salt bridge is formed between Asp9 at the end of the lid and Arg91 at the end of the last helix of the other ATP-cone. Strikingly, the dimer of ATP-cones is slightly offset from the dimer axis of the core protein, such that the ATP-cone of the inactive monomer approaches the core domain of the active monomer, while the ATP-cone of the active monomer is more distant from the core domain of the inactive one (Figure 6a,b). However, the closest atom of any ATP molecule is over 30 Å from the substrate CTP in the active site.

The glycyl radical site is found at Gly711, at the tip of a loop projected from the C-terminal region of the enzyme into the barrel, where it approaches the radical initiator cysteine Cys416 at the tip of a loop stretching through the barrel from the other side (Figure 7).

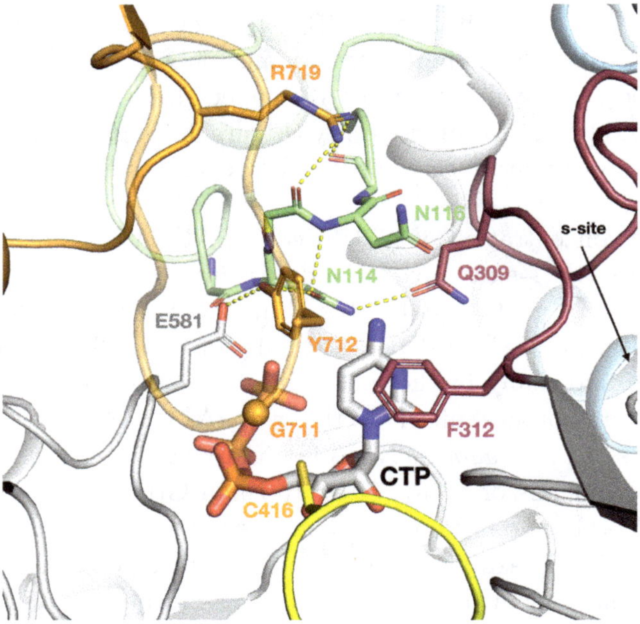

Figure 7: The intricate network of interactions between the N × N flap, C-terminal glycyl radical domain (GRD), loop 2, and substrate in the ATP-CTP complex of PcNrdD. The N × N flap is coloured light green, the GRD orange, loop 2 dark red, and the Cys radical loop yellow. The Gly radical loop is semi-transparent for clarity. The CA atom of Gly711 is indicated by an orange sphere. Important hydrogen bonds are shown as dotted lines.
The online version of this article includes the following figure supplement(s) for figure 7:
Figure supplement 1. The tunnel leading from the active site to the surface of PcNrdD in the ATP-CTP complex.

The second cysteine necessary for the radical mechanism of PcNrdD is Cys171 on the first β-strand of the barrel (Figure 6e). Consistent with the fact that PcNrdD belongs to the group of anaerobic RNRs that use formate as overall reductant (Burnim *et al.*, 2022a; Mulliez *et al.*, 1995; Wei *et al.*, 2014), there is no third cysteine residue on the sixth β-strand, and its place is taken by Asn435. As usual for GREs, Gly711 is completely buried within the barrel, with no solvent-accessible surface area. The base, phosphate, and most of the ribose of substrate CTP are well-defined but the density is weaker around the 5′C-atom (Figure 6e).

After the glycyl radical loop emerges from the α/β barrel, it traverses the top of the barrel, forming a short helix from residues 723 to 730 and ending in an extended tail. This conformation of the C-terminus is very similar to the one seen for TmNrdD in PDB entry 4U3E (Wei *et al.*, 2014). This tail makes few specific interactions with other residues in the core.

When compared to previously determined NrdD structures, the first common structural element is a long α-helix in the core domain stretching from Thr123 to Leu141. Between this and the ATP-cone are the linker (92–110), followed by the flap over the substrate in the active site (111–122). Two residues from the flap make H-bonds to CTP: Asn114 to the γ-phosphate group and Asn116 to the amino group of the cytosine base. We call this region the 'N $\times$ N flap', as the N $\times$ N motif (114–116) is one of the most highly conserved sequence motifs in the NrdD family (Figure 6f). This is the first time such interactions have been observed in an NrdD, as the flap corresponds to a disordered segment of 17 residues in the structures of TmNrdD and T4NrdD. However, an AlphaFold2 (Jumper *et al.*, 2021) model of TmNrdD (entry Q9WYL) suggests that this linker potentially has the same conformation in TmNrdD and possibly all other NrdDs, whether or not they have an ATP-cone. As seen previously in TmNrdD, the triphosphate moiety is recognised through the conserved H $\times$ HD motif (157–160), H-bonds to the main chain atoms of a loop (577–581) following the third-last strand of the α/β barrel and the dipole of the following helix (Figure 6f).

Significantly, the N × N flap forms the nexus of a network of interactions (Figure 7) between the substrate, the linker to the ATP-cone, the C-terminal GRD, and 'loop 2' (residues 304–311) that is responsible for communicating the substrate specificity signal from the s-site at the dimer interface to the active site (Aurelius *et al.*, 2015; Larsson *et al.*, 2001; Logan *et al.*, 1999). Important interactions linking the N × N flap to the C-terminus include H-bonds from Tyr712, which immediately follows the radical Gly711, to Glu581, which bridges to both the substrate and Ala115 in the flap, as well as from Arg719 in the GRD to the main chain carbonyl groups of Asn114 and Met117 in the flap. Tyr712 is highly conserved in the NrdD family (Figures 6f and 7). This nexus is crucial for allosteric activity regulation, as will be discussed later.

As the N × N flap appears to close off the top of the active site, we used the web server Caver Web v1.1 (Stourac *et al.*, 2019) to analyse substrate access to the active site in the ATP-bound form, and identified a tunnel ~12 Å long leading from the surface of the protein to the active site that is 5.6 Å in diameter at its narrowest point (Figure 7 — figure supplement 1). Thus, in principle a fresh substrate could diffuse into the active site on each catalytic cycle.

Cryo-EM structure of the ATP-dTTP-GTP dimer

In order to probe the generality of the conformations observed in the ATP-CTP complex, we also solved the structure of PcNrdD in complex with ATP in the a-site, specificity effector dTTP in the s-site, and substrate GTP in the active site (c-site). Again, 2D classes show a predominantly dimeric form with no more than ~10% tetramers. From a set of 437,886 particles selected from 2D classes with the best apparent density for the ATP-cone region, we obtained a volume at 2.40 Å resolution for the dimer with C1 symmetry that was further improved using DeepEMhancer (Figure 8 — figure supplement 1). dTTP is clearly visible in the s-site (Figure 8 — figure supplement 2a) as is GTP in the active site (Figure 8 — figure supplement 3a). A Mg^{2+} ion can be modelled in the s-site, coordinated by Glu290 and all three phosphate groups

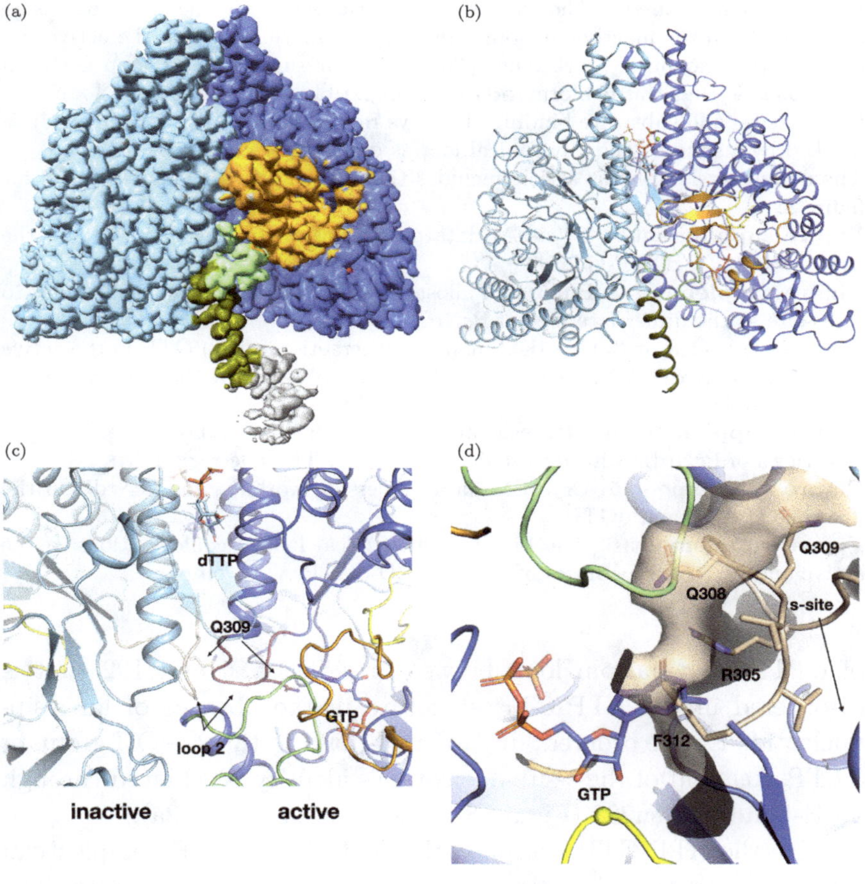

Figure 8: (***Continued***). Structure of PcNrdD in complex with a-site effector ATP, s-site effector dTTP, and substrate GTP. (a) Volume for the PcNrdD dimer in complex with ATP, dTTP, and GTP. The volume is contoured at a low level to emphasise the weak density for the more ordered ATP-cone domain in the active monomer. (b) Ribbon diagram of the ATP-dTTP-GTP complex coloured as in Figure 6b, except that the most N-terminal helix (olive) now extends from residues 85 to 104. (c) Zoom in on panel (b) to illustrate the highly asymmetrical loop 2 conformations in the active and inactive monomers. Loop 2 in the inactive monomer (light blue) is coloured wheat, while in the active monomer (dark blue) it is purple. The conformation of loop 2 that forms a complementary cradle for the guanosine base of the substrate GTP in the active monomer induces a conformation of the other loop 2 that precludes substrate binding. (d) Severe steric clash of loop 2 of the inactive monomer with the substrate and $N \times N$ flap.

Figure 8 (***Continued***): The view is rotated approximately 180° from panels a–c. Loop 2 from the inactive monomer (wheat) is superimposed on the active site of the active monomer (dark blue). The molecular surface of the loop is shown to emphasise that this conformation is incompatible with an ordered $N \times N$ flap (light green) and substrate binding. The Cys radical loop is yellow and Cys416 is marked by a sphere. The Gly radical loop is omitted for clarity.
The online version of this article includes the following figure supplement(s) for figure 8:
Figure supplement 1. Cryo-EM data processing workflow for PcNrdD in the presence of ATP-dTTP-GTP.
Figure supplement 2. Binding of allosteric substrate specificity nucleotides to the s-site at the dimer interface of PcNrdD.
Figure supplement 3. Cryo-EM map and interactions of (a) GTP in the active site of the PcNrdD-ATP-dTTP-GTP complex; (b) ATP in the PcNrdD-ATP-dGTP complex.
Figure supplement 4. Representative two-dimensional (2D) classes showing side views of PcNrdD with diffuse density in the ATP-cone region.
Figure supplement 5. Cryo-EM data processing workflow for PcNrdD in the presence of ATP and dGTP.
Figure supplement 6. The highly asymmetrical loop 2 conformations in the active and inactive monomers of the PcNrdD-dGTP-ATP complex.

of dTTP. A β-hairpin loop between residues 185 and 192 that is disordered in the ATP-CTP structure in the absence of an s-site nucleotide is here ordered, due to an H-bond to the $2' - \text{OH}$ group of dTTP. Readout of the s-site nucleotide's identity is achieved through an H-bond to Asn298 (Figure 8 — figure supplement 2a).

Like the ATP-CTP complex, the ATP-dTTP-GTP complex can also be partitioned into an active and an inactive monomer. The ATP-cone region is very flexible, but the last helix of the ATP-cone in the active monomer can be traced back to residue 85 (Figure 8a, b). The helix is not kinked at residue 91 as in the ATP-CTP dimer but forms an uninterrupted helix between residues 85 and 104. The volume for the ATP-cone region lies entirely on one side of the PcNrdD dimer axis, as is also apparent in some of the 2D classes (Figure 8 — figure supplement 4). The ATP-cone cannot be modelled in detail, but if the volume is contoured at very low level, there is almost enough density for one ATP-cone. Taken together, this suggests that there is one partially disordered ATP-cone and one highly flexible in the ATP-dTTP-GTP complex. In the active monomer, the N × N flap and entire C-terminus are ordered while in the inactive monomer, the entire assembly of linker and N × N flap up to residue 120 and the entire GRD are so flexible that they cannot be seen in the cryo-EM volumes. We expect that the GRD behaves largely as a rigid body whose displacements relative to the core domain become larger in the dATP-bound form.

Intriguingly, binding of dTTP to the s-sites at the dimer interface induces two distinct conformations of loop 2 (Figure 8c). In the active monomer, loop 2 is projected towards the aforementioned network of interactions between N × N flap, GRD, and substrate, packing against the flap, though no polar interactions are apparent. However, loop 2 also makes extensive interactions with its equivalent in the inactive monomer, which forces the latter into a conformation where it would sterically clash with an ordered N × N flap (Figure 8d). Furthermore, Phe412, which stacks on the substrate base in the active monomer, and Arg305, which is projected from the substrate-distal side of loop 2, sterically prevent the substrate binding in the inactive monomer. These may be important contributing factors to the disorder of the N × N flap and GRD. Finally, Gln309, which interacts with the substrate in the active monomer (Figure 8c), is oriented away from the substrate through its interactions with the 'active' loop 2.

Cryo-EM structure of the ATP-dGTP dimer

As a third insight into the ATP-bound forms of PcNrdD, we solved the structure of PcNrdD in complex with specificity effector dGTP in the s-site and ATP at a concentration where it would both act as allosteric effector at the a-site and as the cognate substrate for dGTP (Figure 8 — figure supplement 5). Density for dGTP is clearly visible in the s-site (Figure 8 — figure supplement 2b) and for ATP in the active site (Figure 8 — figure supplement 3b). Again, the two monomers can be divided into an active one and an inactive one. The conformations of loop 2 are once again asymmetrical (Figure 8 — figure supplement 6). The conformation in the active monomer is well-defined, while loop 2 in the inactive monomer is more flexible than in the ATP-dTTP-GTP complex, but still compatible with the hypothesis that the asymmetric arrangement of loop 2 is responsible for disorder of the N × N -GRD-substrate network in the inactive monomer.

Cryo-EM structure of the dATP-bound dimer

In micrographs of PcNrdD samples in the presence of 0.5 mM dATP, we observed an approximately 1:1 mixture of dimeric and tetrameric particles, consistent with results from GEMMA. To resolve whether dATP inhibition was linked to oligomerisation, we resolved the structures of each of these oligomers separately. We made a reconstruction of the dATP-bound PcNrdD dimer from 1,009,021 particles to 2.7 Å resolution that was further improved by DeepEMhancer postprocessing (Figure 9 — figure supplement 1). The core domain is very similar to that of the ATP-CTP complex, with an rmsd in Cα positions of 1.2 Å for 1089 Cα atoms considering each dimer as a whole. dATP is clearly visible in the specificity site (Figure 8 — figure supplement 2).

Strikingly, in contrast to the ATP complexes, the entire GRD is disordered in both monomers of the dATP-bound dimer (Figure 9 — figure supplement 2). No structure is visible after residue 676 at the end of the last strand of the barrel. This suggests that inhibition by binding of dATP to the ATP-cone is coupled to disordering of the C-terminal domain. The N × N flap that covers the top of the substrate in the ATP/CTP-bound form is also disordered, as are the ATP-cones themselves.

Cryo-EM structure of the dATP-bound tetramer

From the same micrographs that revealed the dATP-bound dimers we also resolved dATP-bound tetramers. We obtained a map from cryoSPARC at 2.8 Å, which was again improved with DeepEMhancer (Figure 9a; Figure 9 — figure supplement 1). Two dATP molecules are bound to each ATP-cone and one at each of the specificity sites, giving a total of 12 visible dATPs per tetramer. The tetramer reveals an arrangement of two dimers (chains *A*/*B* and *C*/*D*, respectively)

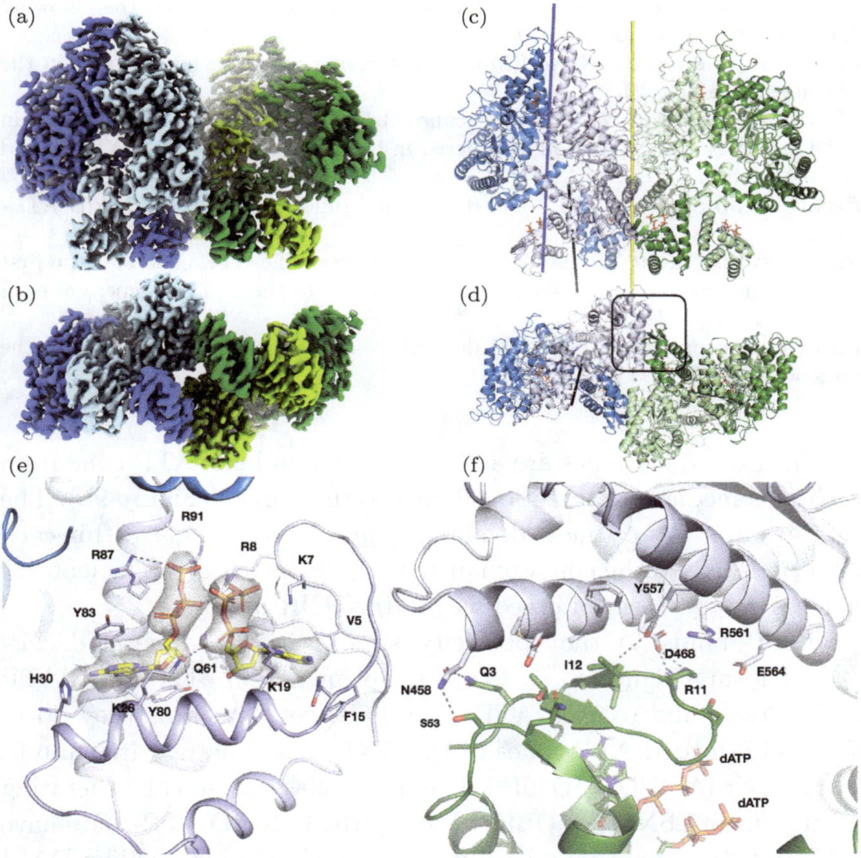

Figure 9: (***Continued***). Structure of the dATP-bound PcNrdD tetramer. (a) Cryo-EM reconstruction coloured by chain: dimer 1 light blue and pale blue; dimer 2 light green and pale green. The four ATP-cones are at the bottom of the figure. (b) View rotated by 90° around a horizontal axis relative to (a) and thus viewed from the bottom, showing the ATP-cones. (c, d) Cartoon representation of the tetramer from the same angles as panels (a) and (b), respectively, and with the same chain colouring. The twofold symmetry axis relating the two dimers of the tetramer is shown as a yellow line. The 12 dATP molecules in the tetramer are shown as sticks. The dimer axis of the left-hand dimer's core domains is shown as a dark blue line and the dimer axis of the ATP-cone pair as a black line. (e) Details of the interaction of dATP molecules at the allosteric activity site in the ATP-cone. (f) Closeup of the interaction area between the roof of the ATP-cone of one dimer and the core domain of the other dimer (marked with a black box in panel d). The most significant residues are labelled.

Figure 9 (***Continued***): The online version of this article includes the following figure supplement(s) for figure 9:
Figure supplement 1. Cryo-EM data processing workflow for PcNrdD in the presence of dATP.
Figure supplement 2. The ATP-cones, linker, entire glycyl radical domain (GRD), and N × N flap are all disordered in both monomers of the dATP-bound PcNrdD dimer.
Figure supplement 3. ATP and dATP bind differently to the PcNrdD ATP-cone.
Figure supplement 4. The linker region between the ATP-cone and the first helix of the core domain interacts very differently in the two monomers of each dimer.
Figure supplement 5. Cryo-EM data processing workflow for PcNrdD in the presence of dATP/CTP.

in which the ATP-cones are all well-ordered and one ATP-cone from each pair mediates interactions between the dimers (Figure 9b). The ATP-cones have an almost domain-swapped arrangement within each dimer relative to the core domains. This oligomeric arrangement has not previously been observed within the RNR family.

dATP binds to the specificity site as in the dimeric dATP-bound form (Figure 8 — figure supplement 2c). The two dATP molecules bound to each ATP-cone do so very much as they do to the dATP-loaded ATP-cones of the RNR class I active site subunit from PaNrdA (PDB ID 5IM3) and the class I radical generating subunit from LbNrdB (5OLK) to which the PcNrdD ATP-cones have 29% and 33% sequence identity, respectively (Table 4). The DALI server reveals an rmsd in Cα positions of 1.9 and 1.7 Å, respectively. The next most similar ATP-cone is that of the RNR transcriptional regulator NrdR from *Streptomyces coelicolor* in its dodecameric form (Rozman Grinberg *et al.*, 2022) binding two ATP molecules, with 82 aligned residues, rmsd 2.4 Å and sequence identity 17%.

The two dATP molecules bind with their negatively charged triphosphate moieties oriented towards each other (Figure 9c). No Mg^{2+} ion is observed between the triphosphate tails, but this may be an artefact of the cryo-EM method, as the PaNrdA and LbNrdB crystal structures suggest that this ion is necessary for

Table 4: Structural similarity comparisons to PcNrdD ATP-cone alone. The PcNrdD ATP-cone in the comparison is the one from the dATP tetramer structure.

PDB ID	Protein	No. of bound nucleotides	Z-score	Residues aligned	RMSD (Å)	Sequence identity (%)
51M3	*P. aeruginosa* NrdA	2 dATP	12.2	92	1.9	29
5OLK	*L. blandensis* NrdB	2 dATP	12.1	93	1.7	33
7P37	*S. coelicolor* NrdR	2 ATP	10.0	82	2.4	17
6AUI	Human NrdA/dATP	1 dATP	9.5	88	2.8	19
7AGJ	*A. aeolicus* NrdA/ATP	2 ATP	8.7	88	2.2	17
5R1R	*E. coli* NrdA E441A	—	8.2	88	3.3	16
7MDI	*N. gonorrhoeae* NrdA	1 dATP	7.4	87	3.4	16

charge neutralisation. As in the ATP-bound form, further charge compensation is achieved by extensive coordination by, or proximity of, basic side chains, for example Lys7, Arg8, Arg87, and Arg91. As previously observed in PaNrdA and LbNrdB, the adenosine base of the 'inner' dATP (dATP1) fits into a pocket under the β-hairpin roof of the ATP-cone and the base of the 'outer' nucleotide (dATP2) is sandwiched between an aromatic residue (Tyr83) and the side of the first α-helix in the domain. No direct contacts are seen between the dATP molecules bound to the ATP-cone and residues in the core domain.

Figure 9 — figure supplement 3 shows the differences in binding modes of dATP in the dATP-bound tetramer and ATP in the ATP-CTP dimer. The high mobility of the ATP-cone in the latter precludes a detailed analysis. However, it is clear that the triphosphate tails have very different orientations. In the ATP complex, the tail of ATP2 is oriented away from the C-terminal helix towards the base of ATP1, such that the β-phosphate moiety may even H -bond to the base of ATP1. The side chain of Arg87 fills the space vacated by the γ-phosphate of ATP2. Instead of H-bonding to the γ-phosphate of ATP2, Arg91 swings away and interacts with the equivalent moiety of ATP2 in the ATP-cone of the other chain. The structures suggest that the additional $2'$-OH group in ATP may cause conformational changes in the ribose ring that propagate to the phosphate tail, but further analysis requires improved resolution.

As in PaNrdA and LbNrdB, the ATP-cones in the dATP-bound tetramer self-interact through a pair of α-helices at the C-terminal end of the ATP-cone. Remarkably, the pair of ATP-cones is highly asymmetrically positioned with respect to the twofold axis of its core dimer, being displaced as a rigid body towards the other dimer (Figure 9b). Both pairs of ATP-cones are displaced in the same way, giving the complex overall C2 symmetry around an axis bisecting the pair of dimers. Consequently, the twofold axes of the ATP-cones and the core dimer are not aligned (Figure 9b). The asymmetric arrangement is associated with two distinct conformations of the linker 93–103 between the fourth helix of the ATP-cone (Figure 9 — figure supplement 4) and the core domain. In chains

A and C, a small kink leads from helix 4 into a short helical segment that packs antiparallel to helix 122–142 in the core domain. In contrast, in chains B and D, the short helix is parallel to helix 122–142 and is joined to helix 4 at a 45° angle. The rest of the polypeptide between residues 103 and 121, including the N × N flap, is disordered in both linkers.

Interestingly, all contacts between the two dimers of the tetramer are mediated by interactions between one of the ATP-cones in one dimer (B or D) and one of the core domains of the other dimer (C or A, respectively). Figure 9d shows the interactions of the roof of the ATP-cone with two outer helices from the C-terminal half of the barrel domain (residues 458–485 and 550–569, preceding the sixth and seventh strands). The interactions are mostly hydrophobic but are reinforced by several H-bonds, for example between Gln3D-Gln458A, Ser53D-Gln458A, Arg11D-Asp468A, the main chain amide of Ile12D and Tyr557A. The amount of buried surface area is small: 717 $Å^2$, or around 1% of the total surface area of each dimer. For comparison, the monomer-monomer interactions within each dimer bury 11.7% of the total area. The residues on the core domain of PcNrdD involved in interactions with the ATP-cone of the other dimer do not show high sequence conservation, even when the alignments are restricted to the most similar sequences.

Cryo-EM structures of PcNrdD-dATP complexes produced in the presence of CTP

To investigate whether substrate had any allosteric effect on the conformations of the ATP-cone, flap, and GRD, we determined separate cryo-EM structures of tetrameric and dimeric PcNrdD, from the same grid, in the presence of 0.5 mM dATP and 0.5 mM CTP, achieving a resolution of 2.6 Å from 1, 105, 348 particles before post-processing for the tetramer and 2.6 Å from 1,132,695 particles for the dimer (Figure 9 — figure supplement 5). Both forms were almost identical to those seen in the presence of dATP alone, and no CTP could be seen bound to the active site (Figure 6 — figure supplement 3). This strongly suggests that substrate binding is dependent on ordering of the flap region by binding of ATP to the ATP-cone.

Hydrogen-deuterium exchange mass spectrometry

To validate our structural and biochemical hypotheses, we further examined the degree of order of the different structural elements of PcNrdD in solution using hydrogen-deuterium exchange mass spectrometry (HDX-MS). This method measures the degree of protection of peptides from H/D exchange in a deuterated solution, which reports on how exposed they are. Three different samples were analysed: apo-PcNrdD and the complexes with dATP-CTP and ATP-CTP.

The apo form shows high deuterium uptake over the entire ATP-cone, linker, and N $\times$ N flap, from residues 1 to 117 (Figure 10a). Analysis of differences in uptake in the N-terminal region between the apo protein and the ATP or dATP complexes clearly validates binding of both nucleotides to the ATP-cone (Figure 10b). Peptides spanning residues 2–84 exhibit clear protection from HDX in both nucleotide-bound states. Furthermore, bimodal deuterium uptake (James *et al.*, 2022) was observed for N-terminal peptides spanning residues 2–23 and 83–99. For peptides very close to the N-terminus (2–23), the apo form only exhibits one high uptake (dynamic) conformational state while for the two ligand bound states there was a clear bimodal behaviour with one high uptake and one protected state of an approximately equal ratio (Figure 10c and Figure 10 — source data 1). In-depth analysis of the causes is difficult, but possible interpretations are that the bimodality is due to a mixture of occupied and empty ATP-cones, or more likely to the mixture of dimers and tetramers.

Interestingly, peptides spanning amino acids 81–99 in the linker region have a bimodal uptake that is most pronounced in the apo state. For the nucleotide-bound forms there are also two populations that converge into a single state at 3000 s (Figure 10c). This suggests a mixture of conformational states with flexible and

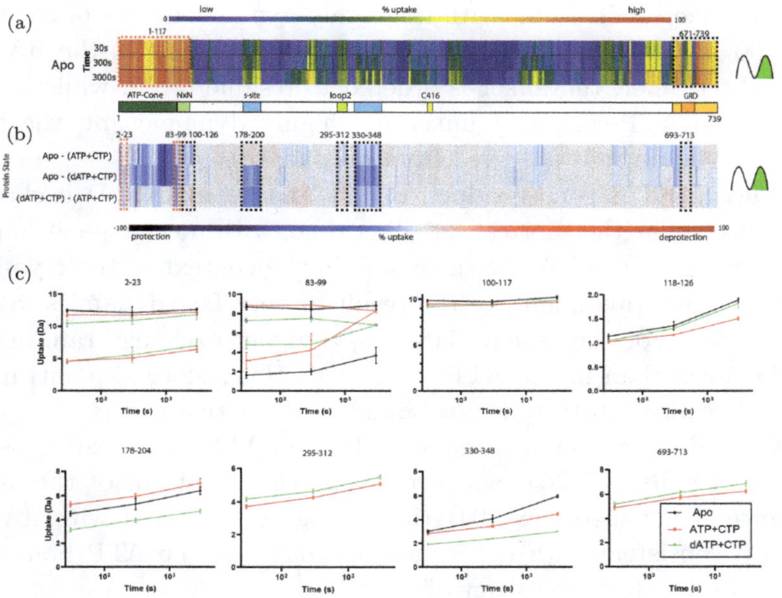

Figure 10: Summary of the hydrogen-deuterium exchange mass spectrometry (HDX-MS) experiments. (a) Heat map showing the deuterium uptake of the apo state for time points 30,300, and 3000 s. Warm colours represent fast uptake and cold colours a low uptake. Both the N- and the C-terminal regions have a high uptake already at 30 s. For the N-terminal bimodal uptake, only the fast-exchanging population (rightmost) is depicted. The influence of bimodal uptake and more detail can be seen in Figure 10 — source data 1. (b) Differential chiclet plot (Zhang *et al.*, 2021) collapses the timepoints into a single maximum deuterium uptake difference for states apo-(ATP + CTP), apo-(dATP + CTP), and (dATP + CTP)-(ATP + CTP). Colours range from deep blue for a high degree of relative protection to deep red for high deprotection. Areas marked with dashed red boxes are peptides showing bimodal uptake, that is two conformational populations present simultaneously, and black dashed boxes denote areas with peptides having EX2 exchange (only one conformational population in the observed timespan). (c) Deuterium uptake plots from dashed areas in (b). For peptides 2–23 and 83–99, when there are two lines with the same colour, the top traces are for the fast-exchanging population and the bottom for a less dynamic/more protected population. Note that for peptides 83–99, the apo state has two populations at all timepoints, while the ATP and dATP converge at 3000 s to one (dynamic) population. For peptides 295–312 and 693–713, the apo curve has been removed for visual clarity.

The online version of this article includes the following source data for figure 10:

Source data 1. Detailed analyses of the hydrogen-deuterium exchange mass spectrometry (HDX-MS) data, including the bimodal uptake distribution in the N-terminal region.

locked linkers, consistent with the highly dynamic nature of residues 85–100 observed in the cryo-EM structures, where in the dATP-bound tetramer, the linker is ordered to residue ∼100, while in the dimer the ATP-cone and linker are highly dynamic and the first ordered structure is observed from residue 120.

The N × N flap at residues 101–117 has a high H/D exchange rate already at the shortest labelling time, with no change at longer times for any state. While surprising in the context of the cryo-EM observations, this might be the result of very fast dynamics in the N × N flap. For the regions 118–126, a higher exchange rate in the dATP form than in the ATP form at the longest time point, most likely because substrate is not bound to the active site.

Clear differences between the ATP and dATP complexes are seen in the regions 178–204, which contains the β-hairpin of the s-site (Figure 10c, panels 178–204) that is highly protected with dATP. This is consistent with the structures that show no ATP bound at the s-site but full occupancy of dATP.

Peptides between 330 and 348, which are in contact with the β-hairpin, are also protected in the dATP complex. The regions 295–312, that is loop 2, are deprotected in the dATP form relative to ATP. This is also in agreement with the cryo-EM structures, in which loop 2 is partially disordered in the dATP complex but ordered in the ATP complex in the monomer with bound CTP.

The HDX-MS results show a high degree of deuterium uptake in the GRD in all forms, indicating a high degree of mobility. This is consistent with the cryo-EM structures, in which the GRD overall has B-factors higher than the core domain (though the glycyl-radical containing loop has similar B-factors to the core, being buried). From the HDX-MS data, it is clear that the GRD is highly dynamic, and while a trend of increased dynamics for some C-terminal peptides in the GRD can be seen for the dATP-CTP state, a detailed analysis is confounded by the limited time resolution and the complexity of the sample, which consists of a mixture of conformations.

Discussion

RNR is an essential enzyme for all free-living organisms. Its complex regulation at the levels of transcription, overall activity and substrate specificity makes it quite unique in Nature (Mathews, 2016; Mathews, 2018). A fundamental understanding of this regulation could for example pave the way for better antibiotics. The ATP-cone is a small, genetically mobile domain frequently found at the N-terminus of RNR catalytic subunits of all major classes, as well as in the bacterial RNR-specific transcription factor NrdR (Rozman Grinberg *et al.*, 2022). Its role in allosteric regulation of class I aerobic RNRs has been extensively studied. In the presence of dATP, the ATP-cone mediates the formation of a surprising variety of different oligomeric complexes that nevertheless have in common that they block the formation of an (observed or inferred) active complex between the catalytic and radical generating subunits that permits proton-coupled electron transfer on each catalytic cycle (reviewed in Martínez-Carranza *et al.*, 2020).

Despite their catalytic subunits being evolutionarily related to those of class I, anaerobic class III RNRs have a fundamentally different activation mechanism that is likely common to all GREs (Backman *et al.*, 2017; Lundahl *et al.*, 2022). Once generated on the catalytic subunit by encounter of the reductase with its radical-SAM family activase, the glycyl radical can catalyse hundreds of cycles of turnover before being exhausted (Torrents *et al.*, 2001). Allosteric inhibition could thus in principle proceed by two mechanisms: by blocking the initial encounter of the reductase and activase, or by preventing subsequent transfer of the radical to the substrate. The first of these hypotheses is intuitively more similar to the steric blocking mechanisms encountered in class I RNRs; the latter is more challenging to understand, as the glycine and cysteine are buried in close proximity in the active site.

In this work, we present the first biochemical, biophysical, and structural characterisation of an anaerobic, class III RNR containing an ATP-cone, providing the first picture of the structural basis for allosteric activity regulation in this large family. We show that the ATP-cone can bind two molecules of ATP or dATP, but not both simultaneously. HDX-MS confirms that binding of either nucleotide reduces the exposure of the ATP-cone to deuterium exchange and supports the biophysical findings of dATP binding to the s-site. As expected, the presence of an ATP-cone confers inhibition of enzyme activity at high dATP concentrations. GEMMA experiments show that PcNrdD exists in a mixture of dimeric and tetrameric states under most experimental conditions studied, with the equilibrium shifted towards dimers in the presence of ATP. Addition of inhibiting concentrations of dATP shifts the equilibrium towards tetramers. The equilibria are not affected by substrate or specificity effector. EPR spectroscopy clearly shows that the glycyl radical is formed even in the presence of dATP. Together, these results show that dATP inhibition in class III RNRs does not occur by blocking the initial encounter of reductase and activase. Instead, our nucleotide-binding results and structures show that dATP inhibition prevents binding of substrates.

The seven cryo-EM structures of PcNrdD in a variety of nucleotide complexes presented here suggest a mechanism in which binding of ATP or dATP modulates activity by affecting the dynamics of a tightly knit network of structural elements including the critical GRD in the C-terminal region of the enzyme. A summary of the conformational states is shown in Figure 11. In the presence of ATP and substrate CTP, the GRD is fully ordered in one monomer and Gly711 is found at its expected position proximal to Cys416, to which it will deliver its radical to initiate each catalytic cycle and accept it back at the end of the cycle. The CTP substrate is bound in this monomer and a conserved but previously unobserved flap containing two Asn residues (N $\times$ N flap) forms over the top of the substrate, making H-bonding interactions with it. Loop 2, which mediates between the substrate specificity site and the active site, is fully ordered. In contrast, in the other monomer, no substrate is bound, loop 2 is disordered and the flap and C-terminus are more dynamic. At the same time, the ATP-cones themselves are relatively flexible. This is not unexpected for this type of ATP-cone,

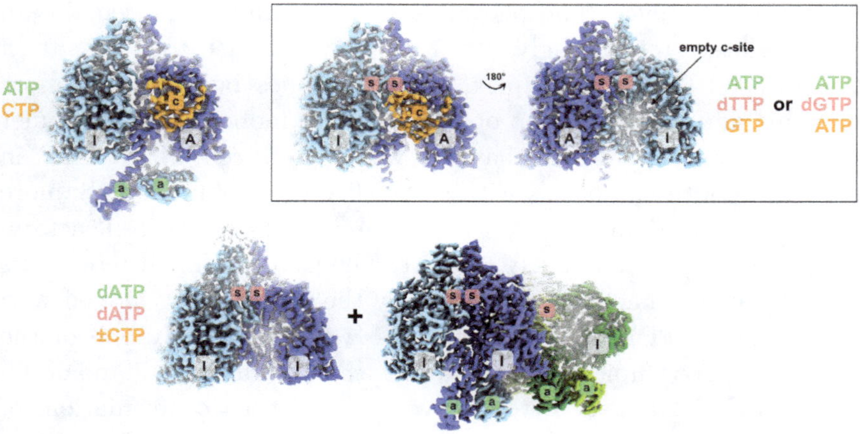

Figure 11: Schematic showing the different structural states of PcNrdD described in this work. The glycyl radical domain (GRD) is coloured orange where present. Inactive and active monomers are labelled I and A, respectively. The a-site in the ATP-cone is labelled with 'a' on a green background, the s-site at the dimer interface with 's' on a red background and the active site with 'c' on an orange background. Labels are only present when a nucleotide is observed in the structure. The ligands bound are listed, coloured according to the sites they bind to. In the dATP/dATP-CTP tetramer, one of the inactive monomers and its s-site are hidden behind the dimer in the foreground.

which binds two nucleotides and was first observed in the class I RNR PaNrdA (Johansson *et al.*, 2016). SAXS studies showed that the ATP-cone in PaNrdA is flexible in the presence of ATP, while in its dATP-bound form it forms a symmetric interaction with an ATP-cone on another dimer, causing the enzyme to tetramerise. A similar behaviour was seen even when this type of ATP-cone was evolutionarily 'transplanted' to the N-terminus of the radical generating subunit, as seen in LbNrdB (Rozman Grinberg *et al.*, 2018a).

By careful isolation of a subset of particles, we were able to characterise an unusual dimer of ATP-cones in the ATP-CTP complex, where four ATP molecules bind with all their triphosphate tails pointing towards each other. Such an orientation is unprecedented for RNRs, but such tail-to-tail orientations, compensated by extensive positive charges, are found in for example adenylate kinases

(Berry *et al.*, 1994). The relevance of this arrangement remains to be elucidated. Interestingly, when dTTP was added to the s-site in the ATP-dTTP-CTP complex, the ATP-cones became even more dynamic and could only be observed in one monomer of the dimer as diffuse density. This behaviour was reproduced with dGTP in the s-site. Both complexes with s-site effector are functionally more relevant forms than the ATP-CTP complex, as in the cell, a form with empty s-site is unlikely to exist. The partitioning of monomers into active and inactive is amplified by the presence of s-site effector (dTTP or dGTP) in both complexes with ATP. Only one of the active sites is occupied by substrate, with ordered N × N flap and C-terminal GRD. This is correlated with two distinct conformations of loop 2 in the two monomers. In the inactive monomer, the N × N flap and GRD are disordered, as loop 2 blocks their active conformation and concomitantly substrate binding. These results point to only one of the PcNrdD monomers being active at a given time and also to a significant cross-talk between the a-site, s-site, and active site. This is corroborated by the finding that an s-site effector is necessary for retention of ATP by the ATP-cone when PcNrdD is first incubated with these nucleotides then purified without them.

In the presence of inhibitory concentrations of dATP we were able to isolate structures of both dimers and tetramers from the same cryo-EM sample, in agreement with the GEMMA results that show a shift towards the tetrameric form. In both forms, the C-terminal GRD is highly mobile in all monomers such that it cannot be seen in the cryo-EM reconstructions. An exposed glycyl radical is incompatible with its stability, so it is possible that the dynamics of the GRD are exaggerated somewhat by the vitrification process. Consistent with this, the HDX-MS results show a small but noticeable deprotection of peptides from the GRD in the presence of dATP-CTP compared to ATP-CTP.

In the tetramers, pairs of ATP-cones asymmetrically disposed relative to the twofold axes of their respective PcNrdD dimers associate through the 'roof' of the 'inner' dATP-binding site to the core domain of the other dimer. While this allowed us to build complete models for the ATP-cones and analyse dATP binding, the role of tetramerisation in the inhibition mechanism is unclear. The tetrameric structures appear to be facilitated by a dATP-induced 'stiffening' of the pair of ATP-cones, similar to what was seen in PaNrdA and LbNrdB, and their interaction with a second dimer of PcNrdD, but the interaction area is small (1% of the total surface area) and the residues involved are not highly conserved in NrdDs, even from the most similar sequences. Therefore, further studies of NrdDs containing ATPcones are required in order to determine whether oligomerisation is a general phenomenon in the family. Furthermore, the GRD is mobile in all complexes with dATP in the a-site, whether dimeric or tetrameric. Since both the ATP-cone and the GRD are disordered in the dATP-bound dimers, the exact molecular mechanism by which dATP induces disorder of the GRD remains somewhat elusive.

Taken together, our results show a clear correlation between ATP or dATP binding to the ATP-cone and activity or inhibition, respectively. The dynamical transitions involve a previously uncharacterised but highly conserved structural element, the $N \times N$ flap, that folds over the substrate and furthermore acts as a structural bridge between the ATP-cone and the C-terminal GRD. An increase in dynamics of these elements prevents substrate binding and thus transfer of the radical from the GRD to the substrate. Interestingly, a similar flap is formed over the top of the active site in the E. coli active complex, but it is formed by part of the C-terminal region of the separate radical generating subunit (Kang *et al.*, 2020). Thus, a locking of the substrate in the active site appears to be a conserved feature of active class I and III RNRs. However, in class I RNR there is no tunnel allowing substrate access to the active site in the locked state. This is consistent with the fact that the class I subunits dissociate on each catalytic cycle, while in class III the glycyl radical can catalyse many substrate reductions before having to be regenerated. It remains enigmatic exactly how the tiny chemical difference between dATP and ATP (the 2′-OH group) results in such major conformational changes, as in the ATP-CTP complex, where the ATP-cones are visible, the a-site in the ATP-cone is separated from the $N \times N$ flap by at least 30 Å and in the most biologically relevant complexes with additional s-site effector, the ATP-cone is very disordered. Nevertheless, the present results give first insights into allosteric activity regulation in anaerobic RNRs and add yet another aspect to the surprisingly wide range of allosteric conformational changes that can be mediated by the binding of the two highly similar nucleotides ATP and dATP to a small, evolutionarily mobile protein domain.

Materials and methods

Cloning of nrdD and nrdG from P. copri

The genes encoding the reductase (NrdD) and activase (NrdG) proteins from *P. copri* were synthesised by GenScript with codon optimisation for *E. coli* and subcloned into the pBG102 plasmid (pET27 derivative) (Center for Structural Biology, Vanderbilt University) between the BamHI and EcoRI restriction sites to produce His6-SUMO-NrdD and His6-SUMO-NrdG protein constructs.

Overexpression and purification of PcNrdD and PcNrdG

Plasmids containing the *nrdD* gene or the *nrdG* gene were transformed into *E. coli* BL21 (DE3) star and *E. coli* BL21 (DE3) competent cells, respectively. Cells (6 I) were grown at 37°C in Luria Broth supplemented with kanamycin (50 μg/ml) to an A_{600} of 1.2. Protein expression was then induced with 0.5 mM isopropyl β-D-1-thiogalactopyranoside (IPTG) and incubation was extended overnight at 20°C. After centrifugation, pellets were suspended in 60 ml of lysis buffer 1 a (50 mM Tris-HCl pH 8, 500 mM KCl, 0.5 mM tris(2-carboxyethyl)phosphine [TCEP]) for PcNrdD and lysis buffer 2a (50 mM Tris-HCl pH 8, 500 mM NaCl, 0.5 mM TCEP) for PcNrdG, supplemented with lysozyme (0.5 mg/ml), and disrupted by sonication. Cell debris was removed by ultracentrifugation at 210,000 $\times g$ for 1 hr at 4°C. The supernatant was then loaded on an immobilised metal affinity Ni-NTA column (HisTrap 5 ml; Cytiva) equilibrated with buffer 1 b(50 mM Tris-HCl, pH 8,300 mM KCl) or buffer 2 b(50 mM Tris-HCl, pH 8,300 mM NaCl), supplemented with

50 mM imidazole as appropriate. The column was washed extensively with the corresponding buffer containing 50 mM imidazole and the His_6-SUMO tagged proteins were eluted using 300 mM imidazole. The proteins were collected and dialysed overnight at 4°C against buffer 1 b or 2 b supplemented with 1 mM 1, 4-dithiothreitol (DTT) in the presence of the PreScission Protease (150 μM) to cleave the affinity-solubility tag. The GST-3C-protease (PreScission) was expressed using pGEX-2T recombinant plasmids. After induction at 25°C with 0.1 mM IPTG for 20 hr, the PreScission Protease was purified using glutathione-Sepharose chromatography. After the dialysis, the cleaved PcNrdD and PcNrdG proteins were centrifuged at 4°C for 10 min and loaded onto a HisTrap column equilibrated with buffer 1 b or 2 b, supplemented with 50 mM imidazole and the flow through containing the cleaved protein was collected. Following cleavage of the His_6-SUMO tag, the proteins incorporated a non-native N-terminal Gly-Pro-Gly-Ser sequence. The purified preparations are called as-purified PcNrdD and PcNrdG, respectively.

A fraction of as-purified PcNrdD was precipitated by addition of HCl, then centrifuged at 15,000 $\times g$ for 5 min. UV-visible spectra of the supernatant were recorded before and after protein precipitation to estimate the amount of nucleotide contamination. PcNrdD preparations with $\leq$ 5% nucleotide contamination are referred to as apo-PcNrdD.

Fractions containing apo-PcNrdD and as-purified PcNrdG were concentrated and loaded on a gel filtration column (Hiload 16/60 Superdex S200; Cytiva) in 25 mM Tris-HCl, pH 8,250 mM NaCl, and 5 mM DTT. The purified protein was concentrated to 20 mg/ml with an Amicon ultrafiltration device (100 kDa cutoff for NrdD and 10 kDa cutoff for NrdG; Millipore), frozen in liquid nitrogen, and stored at −80°C.

[Fe-S] cluster reconstitution and purification of holo-PcNrdG

The reconstitution of the [4Fe–4S] cluster and purification of PcNrdG containing an iron-sulfur cluster (holo-PcNrdG) were performed under strict anaerobic conditions in an Mbraun glove box kept at 18°C and containing less than 0.5 ppm O_2. PcNrdG was treated with 5 mM DTT for 10 min then incubated for 3 hr with a fivefold molar excess of ferrous ammonium sulphate and L-cysteine in the presence of 2 μM *E. coli* cysteine desulfurase CsdA. The holo-PcNrdG was then loaded onto a Superdex 200 10/300 gel filtration column (Cytiva) equilibrated in 25 mM Tris-HCl, 250 mM NaCl, and 5 mM DTT (Figure 1 — figure supplement 2). The peak containing the soluble holo-PcNrdG was then concentrated to 10 mg/ml on a Vivaspin concentrator (10 kDa cutoff).

In vitro enzymatic assay and HPLC analysis

Activity assays were performed under strict anaerobic conditions inside a glovebox in 25 mM Tris pH 8, 50 mM KCl, 10 mM DTT in a volume of 100 μl. A standard pre-reaction mixture contained 1 μM apo-PcNrdD, 2.5 μM holo-PcNrdG, 500 μM S-adenosylmethionine (SAM), 5 mM $MgCl_2$, 1 mM GTP (or 1 mM CTP), and different allosteric effectors dTTP (1 mM), ATP (0–5 mM), or dATP (0–1 mM). The pre-reaction mixture was incubated for 5 min at 37°C before adding simultaneously 10 mM sodium formate and sodium dithionite (12.5 molar excess). When GTP was used as substrate, dTTP was used as s-site effector, and ATP or dATP were titrated to the ATP-cone (a-site). When CTP was used as substrate the s-site was empty at the beginning of the titration and ATP or dATP can plausibly bind both to the s-site and to the a-site in these experiments.

The reaction was incubated for 10 min at 37°C and stopped by the addition of 2.5 μl of 3 M formic acid (FA). Product formation was analysed by HPLC using an Agilent ZORBAX RR StableBond (C18, 4.6 $\times$150 mm, 3.5 μm pore size) equilibrated with buffer A (10% methanol, 50 mM potassium phosphate buffer, pH7, 10 mM tetrabutylammonium hydroxide). Samples of 10 μl were injected and eluted at 0.5 ml/min with a step isocratic flow of 40–100% buffer B (30% methanol, 50 mM potassium phosphate buffer, pH7, 10 mM tetrabutylammonium hydroxide). Compound identification and product quantification were performed by external calibration and NrdD-specific activity was determined.

From a direct plot of activity versus concentration of effector, the K_L values (the concentration of an allosteric effector that gives half maximal enzyme activity) for binding of effectors to the a-site, were calculated in KaleidaGraph using the equation:

$$V = V_{max} \times [NTP/dNTP]/(K_L + [NTP/dNTP]) \quad (7.1)$$

K_i for non-competitive dATP inhibition of GTP reduction was calculated in KaleidaGraph using the equation:

$$V = V_{max}/(1 + [dNTP]/K_i) \quad (7.2)$$

and K_L and K_i for dATP-dependent activation and non-competitive dATP inhibition of CTP reduction were calculated in KaleidaGraph using the equation:

$$v = (V_{max} \times [dNTP]/(K_L + [dNTP])) + (V_{inh}/(1 + [dNTP]/K_i)) \quad (7.3)$$

ITC experiments

ITC experiments were carried out at 20°C on a MicroCal ITC-200 system (Malvern Instruments Ltd) in ITC buffer containing 25 mM 2-[4-(2-hydroxyethyl)piperazin-1-yl]ethanesulfonic acid (HEPES), pH 7.5, 100 mM KCl, 10 mM MgCl, and 0.5 mM TCEP with a stirring speed of 1000 rpm. An initial injection volume was 0.6 μl over a duration of 1.2 s. ApoPcNrdD was prepared in the ITC buffer using SEC, followed by the addition of 1 mM dTTP and 1 mM GTP and incubation at room temperature for 5 min prior to loading to the cell. ATP and dATP ligands were directly diluted in the ITC buffer. Titration of ligand into buffer was also performed to control for heats of dilution and/or buffer mismatch. For dATP-binding analysis, the concentration of PcNrdD in the cell was 50 – 100 μM and dATP concentration in the syringe 0.6 – 1 mM. For ATP titration into PcNrdD, protein concentration in the cell was 200 μM and ATP concentration in the syringe was 0.6 – 1.5 mM. All subsequent injection volumes were 1.7 – 2.5 μ over 3.4 – 5 s with a spacing of 160 s between the injections. Data for the initial injection were not considered. The data were analysed using the one set of sites model of the MicroCal ITC-200 analysis software (Malvern Panalytical). Standard deviations in thermodynamic parameters, N and K_D were estimated from the fits of three different titrations.

Microscale thermophoresis

Both binding of nucleotides to the ATP-cone in PcNrdD and binding of substrates CTP and GTP to PcNrdD were assessed using MST. PcNrdD was labelled using Monolith Protein Labeling Kit RED-NHS 2nd generation according to the manufacturer's protocol. MST was performed using the Monolith NT. 115 instrument (NanoTemper Technologies GmbH) at room temperature. Binding of GTP and CTP was assayed in MST buffer containing 40 mM HEPES pH 7, 50 mM KCl, 50 and 100 mM $MgCl_2$, 5 mM DTT, 0.1% Tween-20, 1 mM dTTP (only for GTP binding), and either 5 mM ATP or 1 mM dATP. The 16 reaction tubes were prepared by the addition of 2 μl MST buffer (concentrated five times), 5 μl GTP or CTP of the desired concentration and 1 μ I RED dye-labelled NrdD in a total volume of 10 μ l. Final NrdD concentration in the binding reaction was 11 nM while the binding partner concentrations were between 70 mM and 2 μM for CTP and 50 mM and 1.5 μM for GTP. Binding of ATP and dATP to the ATP-cone was assayed in MST buffer containing 25 mM HEPES pH 7.5, 100 mM KCl, 10 mM $MgCl_2$, 0.5 mM TCEP, 0.1% Tween-20, 1 mM dTTP, and 5 mM GTP. A 16-step dilution series of the binding partners was prepared by adding 5 μ l buffer to 15 tubes. ATP or dATP was added to the first tube to the final concentration of 4 mM and 5 μl was transferred to the second tube and mixed by pipetting (1:1 dilution series). To each tube of the dilution series 5 μl of PcNrdD was added to reach a final concentration of 11 nM. The samples were transferred to Monolith NT. 115 Series Premium Capillaries (NanoTemper Technologies GmbH), which were scanned using the MST instrument (100% excitation power, medium MST power). Obtained MST data were analysed and fitted using the MO. Affinity Analysis v 2.3 software (NanoTemper Technologies) with default parameters. K_D and standard deviation for GTP and CTP binding to PcNrdD in the presence of ATP and for ATP and dATP binding to the ATP-cone were calculated using fits from at least three individual titrations. K_{D} s for CTP or GTP binding to PcNrdD in the presence of dATP were extremely high and could not be reliably determined since the titration curves did not reach a plateau and therefore are only estimates.

Nucleotide binding to the PcNrdD ATP-cone

PcNrdD was incubated with nucleotides at room temperature for 30 min and then desalted on a NAP-5 column. The desalted protein was boiled for 10 min, centrifuged, and the supernatant was loaded on a Agilent ZORBAX RR StableBond (C18, 4.6×150 mm, 3.5 μm pore size) HPLC column to evaluate the amount of nucleotides retained by the protein. Samples of 100 μl were injected on the column equilibrated with buffer A (10% methanol, 50 mM potassium phosphate buffer, pH 7, 10 mM tetrabutylammonium hydroxide) and eluted at 1 ml/min with a gradient of 40–100% buffer B (30% methanol, 50 mM potassium phosphate buffer, pH 7, 10 mM tetrabutylammonium hydroxide). Compound identification and product quantification were performed by comparison with injected standards. Three different sets of experiments were performed, each set consisting of incubation with only ATP, only dATP, and a combination of ATP and dATP. In the first set of experiments 75 μM apoNrdD was incubated with 1 mM ATP or 1 mM dATP, and when 1 mM each of ATP and dATP was used, dATP was added 15 min after addition of ATP. After 30−min incubation the mixtures were desalted in a buffer without nucleotides. In the second set of experiments 75 μM apo-NrdD in presence of 1 mM dTTP and 5 mM GTP was incubated with 3 mM ATP or 1 mM dATP, and when both nucleotides were combined dATP was added 15 min after addition of ATP. After 30−min incubation the mixtures were desalted in a buffer without nucleotides. In the third set of experiments 75 μM apo-NrdD in presence of 2 mM dTTP and 5 mM GTP was incubated with 3 mM ATP or 1 mM dATP, and when both nucleotides were combined dATP was added 15 min after addition of ATP. After 30–min incubation the mixtures were desalted in a buffer containing 0.1 mM dTTP and 1 mM GTP. It was possible to separate all phosphorylation levels of adenosine and deoxyadenosine nucleotides in the first set of experiments, but not in the second and third sets due to overlapping peaks of the s-site and substrate nucleotides (Table 1).

Gas-phase electrophoretic mobility molecular analysis

The GEMMA instrumental setup and general procedures were as described previously (Kaufman *et al.*, 1996; Rofougaran *et al.*, 2008). Apo-PcNrdD was equilibrated into a buffer containing 100 mM ammonium acetate pH 7.8, then applied onto a Sephadex G-25 chromatography column. In addition, 5 mM DTT was added to the PcNrdD protein solution to increase the protein stability. Prior to analysis, the protein sample was diluted to 2 μM in a buffer containing 100 mM ammonium acetate, pH 7.8, 0.005% (vol/vol) Tween 20, magnesium acetate (equimolar to the total nucleotide concentration) and the corresponding nucleotides (where indicated). The protein samples were then incubated for 2 min at room temperature, centrifuged and applied to the GEMMA instrument for data collection. Each sample was scanned five times and the signal was added together to obtain the traces presented. The experiments were performed with a flow rate driven by 2 psi to minimise noise signals that may appear with elevated nucleotide or protein concentrations at the manufacturer's recommended flow rate driven by 3.7 psi.

Glycyl radical characterisation by EPR

EPR samples were handled under strictly anaerobic conditions inside a glovebox in 25 mM Tris-HCl pH 8, 50 mM KCl, 5 mM DTT in a final volume of 100 μl. A standard pre-reaction mixture containing 75 μM holo-PcNrdG, 1 mM S-adenosyl-methionine, was treated with a 12.5 excess of dithionite and was incubated for 20 min at 37°C. The pre-mixture was added to a mixture of 50 μM apo-PcNrdD, 10 mM sodium formate, supplemented or not with 1.5 mM ATP or 1 mM dATP and in the presence of 1 mM CTP. The final reaction was incubated for 20 min at 37°C and 100 μl sample was transferred into a standard EPR tube (Wilmad LabGlass 707-SQ-250M) and stored in liquid nitrogen before recording the EPR spectrum.

X-band CW-EPR spectra of PcNrdD were recorded on a Bruker ELEXYS-E500 spectrometer operating at microwave frequency of 9.38 GHz, equipped with a SuperX EPR049 microwave bridge and a cylindrical TE_{011} ER 4122SHQE cavity in connection with an Oxford Instruments continuous flow cryostat. Measuring temperature at 30 K was achieved using liquid helium flow through an ITC 503 temperature controller (Oxford Instruments). For achieving non-saturating conditions, a microwave power of $10\,\mu$W was applied for all measurements. Other EPR settings were: modulation frequency of 100 kHz; modulation amplitude of 8 G; and accumulation of 16 scans for an optimal S/N ratio. Double integrated spectra were used for intensity analysis.

Cryo-EM sample preparation and data acquisition

The proteins for cryo-EM analysis were all prepared in the same manner. Apo-PcNrdD was incubated with 1 mM NTP or dNTP at room temperature for 5 min, then centrifuged before injection onto a Superdex S200 10/300 column pre-equilibrated with a buffer containing 25 mM HEPES-NaOH pH 7.5, 100 mM KCl, 0.5 mM TCEP, 10 mM MgCl, and supplemented with varying combinations of 0.5 mM NTP or 0.2 mM dNTP. The relevant peak was collected, then diluted at 0.5 mg/ml and the corresponding nucleotides were added to different final concentrations (0.5–5 mM NTP or 0.5–1 mM dNTP) (Table 5).

The grids were glow-discharged for 60 s at 20 mA using a GloQube (Quorum) instrument, then prepared in the following manner: $3\,\mu$l of sample at 0.5 mg/ml were applied on Quantifoil 1.2, 1.3 300 mesh Cu grids or Quantifoil 2.1, 300 mesh Au grids and plunge-frozen in liquid nitrogen-cooled liquid ethane using a Vitrobot Mark IV (Thermo Fisher Scientific) with a blot force of 1 and followed by 5s blot time, at 4°C and 100% humidity. The grids used for each sample are specified in Table 5.

Table 5: Data collection, processing, and refinement statistics for the cryo-EM structures of PcNrdD.

Ligands	**dATP tetramer**	**dATP dimer**	**ATP-dTTP-GTP dimer**	**ATP-dGTP dimer**	**ATP-CTP dimer**	**dATP-CTP tetramer**	**dATP-CTP dimer**
PDB entry	8P28	8P27	8P2S	8P39	8P23	8P2C	8P2D
EMDB entry	EMD-17359	EMD-17358	EMD-17373	EMD-17385	EMD-17357	EMD-17360	EMD-17361
Concentrations (mM)	0.5		5.0/1.0/1.0	5.0/1.0	0.5/0.5	0.5	
Grids	Quantifoil 1.2/1.3, 300 mesh Cu		Quantifoil 2.1, 300 mesh Au	Quantifoil 2.1, 300 mesh Au	Quantifoil 2.1, 300 mesh Cu	Quantifoil 2.1, 300 mesh Cu	
Pixel size (Å)	0.8676		0.824	0.860	0.8676	0.8464	
Dose rate (e^{-}/px/s)	16.1		15.0	15.0	17.1	13.7	
Exposure time (s)	2.0		2.0	2.0	2.1	2.0	
Total dose (e^{-}/Å^2)	46		40	40	48	38	
Defocus range (μM)	−1.0 to −3.4		−0.8 to −2.2	−0.6 to −2.4	−0.6 to −2.4	−0.6 to −2.0	
Micrographs used (collected)	16,667 (21,804)		4964 (12,501)	14,346 (16,745)	17,033 (21,512)	11,780 (15,268)	8796 (15,268)
Particles in final class	1,349,133	1,009,021	437,866	589,345	291,231	1,105,348	1,132,695
Symmetry	C2	C1	C1	C1	C1	C2	C1
Resolution (FSC = 0.143; Å A)	2.77	2.73	2.40	2.56	3.17	2.59	2.59
Map sharpening *B*-factor (Å^2)	138.5	128.0	70.5	72.3	152.8	107.8	108.9
CC (mask) from phenix.refine	0.76	0.77	0.76	0.63	0.74	0.81	0.64

(*Continued*)

Table 5: (Continued)

Model composition							
Non-hydrogen atoms (residues)	21,141 (2584)	8710 (1089)	9694 (1201)	9705 (1200)	11,399 (1401)	21,137 (2584)	8793 (1087)
Ligands	8 dATP (a) 4 dATP (s)	2 dATP (s)	2 dTTP(s) 1 GTP (c)	2 dGTP (s) 1 ATP (c)	4 ATP (a) 1 CTP (c)	8 dATP (a) 4 dATP (s)	2 dATP (s)
min/mean/max *B*-factors protein (Å^2)	52.8/63.4/84.7	40.2/50.6/66.6	25.6/35.4/72.9	12.6/41.2/98.7	15.7/66.1/124.4	15.5/49.4/136.4	30.9/43.6/78.1
min/mean/max *B*-factors ligands (Å^2)	55.8/62.9/74.5	47.1/47.2/47.4	25.4/33.1/79.3	10.8/33.1/88.0	26.0/81.2/123.7	18.0/46.8/92.7	35.9/36.2/47.2
Deviations from ideal geometry							
rmsd (bonds)	0.005	0.004	0.003	0.003	0.004	0.004	0.002
rmsd (angles)	0.61	0.55	0.54	0.60	0.58	0.59	0.51
Ramachandran plot (%)							
Favoured	90.1	91.7	95.2	93.8	93.8	94.1	92.0
Allowed	9.9	8.3	4.7	5.8	5.9	5.9	8.0
Outliers	0.0	0.0	0.1	0.4	0.3	0.0	0.0
Rotamer outliers	1.9	5.4	3.0	6.0	3.8	4.2	4.8
MolProbity score	2.1	2.4	2.2	2.5	2.2	2.3	2.5
MolProbity clash score	6.5	6.2	8.7	10.0	5.5	7.1	10.2

Grids were clipped and loaded into a 300−kV Titan Krios G2 microscope (Thermo Fisher Scientific, EPU 2.8.1 software) equipped with a Gatan BioQuantum energy filter and a K3 Summit direct electron detector (AMETEK). Grids were screened for quality based on particle distribution and density, and images from the best grid were recorded. Micrographs were recorded at a nominal magnification of 105,000. Details of the other data collection parameters used for each sample are given in Table 5.

Cryo-EM data processing

Data processing was performed in cryoSPARC (Punjani *et al.*, 2017). The first steps in processing of all datasets were patch motion correction and patch contrast transfer function (CTF) estimation, followed by curation of exposures based on ice thickness, poor defocus estimation, etc. Other steps are as detailed below. Resolution estimation in cryoSPARC was done using a gold standard Fourier shell correlation value of 0.143 (Rosenthal and Henderson, 2003). Map post-processing was done using DeepEMhancer (Sanchez-Garcia *et al.*, 2021) and post-processing resolution estimation was performed using DeepRes (Ramírez-Aportela *et al.*, 2019). Models were placed in the maps using either phenix.dock_in_map (Liebschner *et al.*, 2019) or Molrep (Vagin and Teplyakov, 1997) in the CCP-EM package (Burnley *et al.*, 2017). Model building was done by alternating rounds of manual building in Coot (Emsley *et al.*, 2010) with real space refinement in phenix.refine (Liebschner *et al.*, 2019). Secondary structure restraints were used throughout. The inclusion of riding hydrogen atoms reduced the number of bad contacts in all models. An AlphaFold2 (Jumper *et al.*, 2021) model of PcNrdD was used in the later stages of building the first model to resolve some ambiguous loops. All final models have good correlations between map and model and good stereochemical properties (Table 5). All structures have been deposited in the Protein Databank and the corresponding volumes deposited in the Electron Microscopy Data Bank with the accession numbers listed in Table 5.

PcNrdD-ATP-CTP complex

7.0.0.0.1 Data processing

A total of 17,033 movies were used after curation. A low-pass filtered volume was made in Chimera (Pettersen *et al.*, 2021) from the crystal structure of NrdD from *T. maritima* (TmNrdD, PDB ID 4COI). From this volume, a set of templates was created and used for template-based picking with a diameter of 100 Å. Due to the large number of micrographs, they were split into two sets. The largest set (16,219 micrographs) was analysed first. About 8.6 million particles were extracted with a 350-pixel box size and classified into 60 2D classes. Eight representative classes containing different orientations (~3.3 million particles) were used to generate four ab initio models without symmetry (C1) and particles were subjected to 3D heterogeneous refinement using four classes. The most populated model containing ~1 million particles was subjected to homogeneous and non-uniform refinement, which gave a map with an overall FSC resolution of 2.94 Å. A further 435,695 particles were then extracted from the remaining 810 micrographs. These were subjected to 2D classification, and the resulting classes were merged with the larger set, giving 1,199,575 particles. The combined particles were subjected to homogeneous, non-uniform and local refinement with a predefined mask, which gave a map at 2.87 Å (Figure 6 — figure supplement 1).

7.0.0.0.2 Generation of map with better ATP-cone density

The particles were further classified into 50 2D classes, from which 23 classes having 1,040,266 particles were selected. Four ab initio models were generated and 570,730 particles with slightly better ATP-cone density were refined without symmetry. The particles were then subjected to 3D classification using ten classes of ~57,000 particles each. The classes having slightly better ATP-cone density were selected for further processing and final refinement without symmetry gave a map with an overall FSC resolution of 3.17 Å from 291,231 particles, which was post-processed using DeepEMhancer (Figure 6 — figure supplement 1).

7.0.0.0.3 Model building and refinement

A partially complete model for one dimer of the dATP complex (see below) was placed in the map using phenix.dock_in_map. The ATP-cones were built based on the ones from the dATP-only tetramers (see below).

PcNrdD-dATP complexes

7.0.0.0.4 Tetramers

Data processing for first part
A set of 11,161 movies was used after curation, template-based picking based on the structure of TmNrdD and particle extraction using a box size of 448 pixels gave 5, 331, 420 particles. Seven 2D classes having 1,171,839 particles were used to make ab initio models, followed by heterogeneous refinement. The best class having 461,020 particles and volume was used for homogeneous refinement, which gave a map at 3.3 Å. Non-uniform refinement followed by local refinement with a predetermined mask gave a map at 3.1 Å with no imposed symmetry.

Data processing for second part
A set of 16,667 movies was used after curation. Particles were picked and extracted using the same box size as in the first part, which gave 7, 889, 234 particles. Ten 2D classes having 1, 881, 057 particles were used to make two ab initio models, followed by heterogeneous refinement. The best class with 1,083,657 particles was selected and subjected to homogeneous refinement followed by non-uniform refinement and a local new refinement with a pre-determined mask, which gave a map at 2.9 Å with no imposed symmetry.

Merging of datasets and CTF refinement
The best particles from the first dataset were classified into 80 2D classes from which 36 having 399,523 particles were selected. The particles from the second dataset resulting in the best volume were classified into 100 2D classes, of which 31 having 949,610 particles were selected. These particle sets were merged (giving 1,349,133 particles) and refined against the best 3D volume which gave a map

at 2.78 Å. Iterative refinement and NU refinement jobs, ultimately using C2 symmetry, and local CTF refinement with a defocus search range of ±2000, gave a map at 2.8 Å with C2 symmetry that was post-processed using DeepEMhancer (Figure 9 — figure supplement 1).

Model building and refinement

Model building was initially done for one dimer by fitting a homology model generated using SwissModel with T4NrdD as a template to the map using phenix.dock_in_map. Similar homology models for the four ATP-cones were placed by hand into the density followed by rounds of real space refinement. After almost completely rebuilding one dimer, the second dimer was placed into the density and model building and refinement were continued.

Dimers

The first part of the data processing was shared with the PcNrdD-dATP tetramers. In the same micrographs there were a number of 2D class averages that represented dimers. Eight such class averages having 1, 916, 387 particles were used to make two ab initio models. This was followed by rounds of heterogeneous refinement from where the best model having 1,009,021 particles was refined with no symmetry to give a map at 2.8 Å. A final non-uniform refinement with C1 symmetry gave a map at 2.6 Å that was post-processed using DeepEMhancer (Figure 9 — figure supplement 1).

PcNrdD-dATP-CTP complexes

7.0.0.0.5 Tetramers

Blob picking was carried out from 11,780 curated micrographs with a maximum diameter of 200 Å and a minimum diameter of 90. In total 4, 646, 479 particles were then extracted with a box size of 448 pixels. This was followed by 2D classification, ab initio model generation and heterogeneous refinements. The best class with 1,105,348 particles was used for non-uniform refinement with C2 symmetry, which gave a final map at 2.59 Å resolution that was post-processed using DeepEMhancer (Figure 9 — figure supplement 5).

7.0.0.0.6 Dimers

The initial data processing steps were the same as above. Template-based picking from 8796 curated micrographs using a low-pass filtered volume of the dATP-only tetramer gave 6,836,668 particles after extraction with a box size of 300 pixels. This was followed by 2D classifications, selection of classes representing dimers, ab initio model generations and refinement of the best class. The final round of non-uniform refinement with 1,132,695 particles with C1 symmetry gave a map of 2.53 Å resolution. The estimated median resolution after post-processing was 2.1 Å (Figure 9 — figure supplement 5). Model building used the coordinates of the dATP-only dimer as a starting model.

PcNrdD-ATP-dTTP-GTP complex

7.0.0.0.7 Data processing

A total of 12,501 movies were processed and 6105 movies were used after curation. Template-based picking was used with a diameter of 150 pixels. About 3.9 million particles were extracted from 5952 movies with a 400 pixel box size and used for 2D classification into 100 classes. Eleven representative 2D classes containing different orientations (∼2.4 million particles) were used to generate an ab initio model and particles were subjected to 3D heterogeneous refinement using ten classes, resulting in the most populated model containing 437,866 particles. This model was subjected to homogeneous and non-uniform refinement, which gave a map with an overall FSC resolution of 2.47 Å. The particles were then subjected to 3D classification and the resulting classes were merged for further processing and final refinement without symmetry (C1) gave a map with an overall FSC resolution of 2.4 Å that was post-processed using DeepEMhancer (Figure 8 — figure supplement 1).

7.0.0.0.8 Model building and refinement:

A complete model for the dimer of the ATP–CTP structure (see above) was placed in the map using phenix.dock_in_map. The ATP-cones were removed as they were not visible in the reconstruction. After model building and refinement, the final map-to-model correlation value was 0.74. Model quality statistics are presented in Table 5.

PcNrdD–ATP–dGTP complex

7.0.0.0.9 Data processing

Template-based picking was carried out on 14,373 curated micrographs and 4,522,028 particles were extracted with a box size of 350 pixels. This was followed by multiple rounds of 2D classifications, ab initio model generation and refinements. C1 symmetry was applied on the best volume which gave a map with 589,345 particles having a final resolution of 2.58 Å. Model building and refinement used the PcNrdD–ATP/CTP complex as a starting model (Figure 8 — figure supplement 5).

Bioinformatics

To construct an HMM logo of NrdD sequences, a representative selection of NrdD sequences from all domains of life including viruses were collected and aligned with Clustal Omega (Sievers *et al.*, 2011). Subsequently, an HMM model was built with hmmbuild from the HMMER suite (Eddy, 2011) and a logo was created using the Skylign web service using the 'Information Content — Above Background' option (Wheeler *et al.*, 2014). Conserved parts of the logo were extracted and displayed here.

Hydrogen–deuterium exchange mass spectrometry

All chemicals were from Sigma-Aldrich. pH measurements were made using a SevenCompact pH-meter equipped with an InLab Micro electrode (Mettler-Toledo). A 4-point calibration (pH 2, 4, 7, 10) was made prior to all measurements. The HDX-MS analysis was made using automated sample preparation on a LEAP H/D-X PAL platform interfaced to an liquid chromatography-mass spectrometry (LC–MS) system, comprising an Ultimate 3000 micro-LC coupled to an Orbitrap Q Exactive Plus MS.

HDX was performed on 2 mg/ml PcNrdD with and without ligands (5 mM ATP +5 mM CTP or 1 mM dATP + 5 mM CTP) in 25 mM HEPES, 100 mM KCl, 20 mM $MgCl_2$, and 0.5 mM TCEP, pH 7.5. The HDX-MS was performed in one continuous run, the apo state being run before both complexed states. For each labelling time point, 3 μl HDX samples were diluted with 27 μl labelling buffer (containing the ligands) of the same composition prepared in $D_2O, pH_{(read)}$ 7.4. The HDX labelling was carried out at $t = 0, 30, 300,$ and 3000 s at 20°C. Each time point was run in four to six replicates. The labelling reaction was quenched by dilution of 30 μl labelled sample with 30 μl of 1% trifluoroacetic acid, 0.4 M TCEP, 4 M urea, pH 2.5 at 1°C.50 μl of the quenched sample was directly injected and subjected to online pepsin digestion at 4°C on an in-house immobilised pepsin column (2.1 × 30 mm). The online digestion and trapping were performed for 4 min using a flow of 50 μl/min 0.1% FA, pH 2.5. The peptides generated by pepsin digestion were subjected to online solid phase extraction (SPE) on a PepMap300 C18 trap column (1 × 15 mm) and washed with 0.1% FA for 60 s. Thereafter, the trap column was switched in-line with a reversed-phase analytical column (Hypersil GOLD, particle size 1.9 μm, 1 × 50 mm). The mobile phases were (A) 0.1% FA and 95% acetonitrile/0.1% FA (B) and separation was performed at 1°C using a gradient of 5 − 50% B over 8 min and then from 50 to 90% B for 5 min. Following the separation, the trap and column were equilibrated at 5% organic content until the next injection. The needle port and sample loop were cleaned three times after each injection with mobile phase 5% MeOH/0.1% FA, followed by 90% MeOH/0.1% FA and a final wash of 5% MeOH/0.1% FA. After each sample and blank injection, the pepsin column was washed by injecting 90 μl of pepsin wash solution 1% FA/4 M urea/5% MeOH. In order to minimise carry-over, a full blank was run between each sample injection. Separated peptides were analysed on a Q Exactive Plus MS, equipped with a HESI source operated at a capillary temperature of 250°C with sheath gas 12, Aux gas 2, and sweep gas 1 (au). For HDX analysis, MS full scan spectra were acquired at 70 K resolution, AGC 3e6, max IT 200 ms and scan range 300–2000. For identification of generated peptides, separate non-deuterated samples were analysed using data-dependent MS/MS with HCD fragmentation. A summary of the HDX experimental detail is reported in Figure 10 — source data 1. The mass spectrometry raw files have been deposited at the ProteomeXchange Consortium via the PRIDE partner repository (Perez-Riverol *et al.*, 2022) with the dataset identifier PXD047943.

Data analysis

PEAKS Studio X Bioinformatics Solutions Inc (BSI, Waterloo, Canada) was used for peptide identification after pepsin digestion of non-deuterated samples. The search was done on a FASTA file with only the NrdD sequence. The search criteria were a mass error tolerance of 15 ppm and a fragment mass error tolerance of 0.05 Da, allowing for fully unspecific cleavage by pepsin. Peptides identified by PEAKS with a peptide score value of log $p > 25$ and no modifications were used to generate a peptide list containing peptide sequence, charge state, and retention time for the HDX analysis. HDX data analysis and visualisation were performed using HDExaminer, version 3.3 (Sierra Analytics Inc, Modesto, US). The analysis was made on the best charge state for each peptide, allowed only for EX2 (except for bimodal parts of the protein for which EX1 calculation of uptake was allowed) and the two first residues of a peptide were assumed unable to hold deuteration. Due to the comparative nature of the measurements, the deuterium incorporation levels for the peptic peptides were derived from the observed relative mass difference between the deuterated and non-deuterated peptides without back-exchange correction using a fully deuterated sample (Engen and Wales, 2015). As a full deuteration experiment was not made, full deuteration was set to 75% of maximum theoretical uptake. The presented deuteration data are the average of all high and medium confidence results. The allowed retention time window was ± 0.5 min. The spectra for all time points were manually inspected; low scoring peptides, obvious outliers, and any peptides where retention time correction could not be made consistent were removed. As bottom-up labelling HDX-MS is limited in structural resolution by the degree of overlap of the peptides generated by pepsin digestion, the peptide map overlap is shown for the respective state in Figure 10 — source data 1.

Materials availability statement

The plasmids for expression of PcNrdD and PcNrdG are available from the authors upon request.

Acknowledgements

We dedicate this article to PhD candidate Ipsita Banerjee¶ who made seminal contributions to the data processing parts of this study but who passed away suddenly on 7 December 2022. The authors thank Gustav Berggren, Uppsala University, Martin Högbom, Stockholm University, and Marc Fontecave, Collège de France, for letting us use their anaerobic chambers, Anders Hofer, UmeåUniversity, for the GEMMA instrument, David Drew and Henrietta Nielsen, Stockholm University, for the ITC and MST instruments, Anders Olsson, SciLifeLab Stockholm University, for the ITC and HPLC instruments, Annette Roos, SciLifeLab Uppsala University, for her help during the procedure of PcNrdD labelling for MST, Alexey Pisarev and Thomas Jonsen, Agilent, for fixing our HPLC instrument, and Malvern Panalytical for kindly sharing the MicroCal PEAQ-ITC analysis software for the analysis of ITC data. We would like to also thank master student Sina Becker for SEC experiments. Cryo-EM sample screening, optimisation, and data collection were performed at the Cryo-EM Swedish National Facility, funded by the Knut and Alice Wallenberg, Family Erling Persson and Kempe Foundations, SciLifeLab, Stockholm University, and UmeåUniversity. The authors would like to thank Marta Carroni, Karin Walldén, Julian Conrad, Terezia Kovalova, and Victor Tobiasson for their assistance during the cryo-EM experiments. Support from the Swedish National Infrastructure for Biological Mass Spectrometry (BioMS) and the SciLifeLab, Integrated Structural Biology platform is gratefully acknowledged.

This study was supported by grants from the Swedish Research Council (2019-01400 to BMS, 201604855 to DTL), the Swedish Cancer Society (CAN 20 1210 PjF to BMS), and the Wenner-Gren Foundations (to BMS).

References

Andersson J, Westman M, Hofer A, Sjöberg BM. 2000. Allosteric regulation of the class III anaerobic ribonucleotide reductase from bacteriophage T4. *The Journal of Biological Chemistry* **275**: 19443–19448. DOI: https://doi.org/10.1074/jbc.M001490200, PMID: 10748029

Ando N, Brignole EJ, Zimanyi CM, Funk MA, Yokoyama K, Asturias FJ, Stubbe J, Drennan CL. 2011. Structural interconversions modulate activity of *Escherichia coli* ribonucleotide reductase. *PNAS* **108**: 21046–21051. DOI: https://doi.org/10.1073/pnas.1112715108, PMID: 22160671

Ando N, Li H, Brignole EJ, Thompson S, McLaughlin MI, Page JE, Asturias FJ, Stubbe J, Drennan CL. 2016. Allosteric Inhibition of Human Ribonucleotide Reductase by dATP Entails the Stabilization of a Hexamer. *Biochemistry* **55**: 373–381. DOI: https://doi.org/10.1021/acs.biochem.5b01207, PMID: 26727048

Aravind L, Wolf YI, Koonin EV. 2000. The ATP-cone: an evolutionarily mobile, ATP-binding regulatory domain. *Journal of Molecular Microbiology and Biotechnology* **2**: 191–194 PMID: 10939243.

Aurelius O, Johansson R, Bågenholm V, Lundin D, Tholander F, Balhuizen A, Beck T, Sahlin M, Sjöberg B-M, Mulliez E, Logan DT. 2015. The crystal structure of thermotoga maritima class III ribonucleotide reductase lacks a radical cysteine pre-positioned in the active site. *PLOS ONE* **10**: e0128199. DOI: https://doi.org/10.1371/journal.pone.0128199 , PMID: 26147435

Backman LRF, Funk MA, Dawson CD, Drennan CL. 2017. New tricks for the glycyl radical enzyme family. *Critical Reviews in Biochemistry and Molecular Biology* **52**: 674–695. DOI: https://doi.org/10.1080/10409238.2017.1373741, PMID: 28901199

Berry MB, Meador B, Bilderback T, Liang P, Glaser M, Phillips GN. 1994. The closed conformation of a highly flexible protein: the structure of *E. coli* adenylate kinase with bound AMP and AMPPNP. *Proteins* **19**: 183–198. DOI: https://doi.org/10.1002/prot.340190304

Birgander PL, Kasrayan A, Sjöberg BM. 2004. Mutant R1 Proteins from *Escherichia coli* Class la Ribonucleotide Reductase with Altered Responses to dATP Inhibition. *Journal of Biological Chemistry* **279**: 14496–14501. DOI: https://doi.org/10.1074/jbc.M310142200

Brignole EJ, Tsai KL, Chittuluru J, Li H, Aye Y, Penczek PA, Stubbe J, Drennan CL, Asturias F. 2018. 3.3-Å resolution cryo-EM structure of human ribonucleotide reductase with substrate and allosteric regulators bound. *eLife* **7**: e31502. DOI: https://doi.org/10.7554/eLife.31502, PMID: 29460780

Burnim AA, Spence MA, Xu D, Jackson CJ, Ando N. 2022a. Comprehensive phylogenetic analysis of the ribonucleotide reductase family reveals an ancestral clade. *eLife* **11**: e79790. DOI: https://doi.org/10.7554/eLife.79790, PMID: 36047668

Burnim AA, Xu D, Spence MA, Jackson CJ, Ando N. 2022b. Analysis of insertions and extensions in the functional evolution of the ribonucleotide reductase family. *Protein Science* **31**: e4483. DOI: https://doi.org/10.1002/pro.4483, PMID: 36307939

Burnley T, Palmer CM, Winn M. 2017. Recent developments in the CCP-EM software suite. *Acta Crystallographica. Section D, Structural Biology* **73**: 469–477. DOI: https://doi.org/10.1107/S2059798317007859, PMID: 28580908

Crona M, Avesson L, Sahlin M, Lundin D, Hinas A, Klose R, Söderbom F, Sjöberg BM. 2013. A rare combination of ribonucleotide reductases in the social amoeba *Dictyostelium* discoideum. *Journal of Biological Chemistry* **288**: 8198–8208. DOI: https://doi.org/10.1074/jbc.M112.442434, PMID: 23372162

Eddy SR. 2011. Accelerated Profile HMM Searches. *PLOS Computational Biology* **7**: e1002195. DOI: https://doi.org/10.1371/journal.pcbi.1002195, PMID: 22039361

Eliasson R, Pontis E, Sun X, Reichard P. 1994. Allosteric control of the substrate specificity of the anaerobic ribonucleotide reductase from *Escherichia coli*. *The Journal of Biological Chemistry* **269**: 26052–26057. DOI: https://doi.org/10.1016/S0021-9258(18)47158-X, PMID: 7929317

Emsley P, Lohkamp B, Scott WG, Cowtan K. 2010. Features and development of Coot. *Acta Crystallographica. Section D, Biological Crystallography* **66**: 486–501. DOI: https://doi.org/10.1107/S0907444910007493, PMID: 20383002

Engen JR, Wales TE. 2015. Analytical aspects of hydrogen exchange mass spectrometry. *Annual Review of Analytical Chemistry* **8**: 127–148. DOI: https://doi.org/10.1146/annurev-anchem-062011-143113, PMID: 26048552

Fairman JW, Wijerathna SR, Ahmad MF, Xu H, Nakano R, Jha S, Prendergast J, Welin RM, Flodin S, Roos A, Nordlund P, Li Z, Walz T, Dealwis CG. 2011. Structural basis for allosteric regulation of human ribonucleotide reductase by nucleotide-induced oligomerization. *Nature Structural & Molecular Biology* **18**: 316–322. DOI: https://doi.org/10.1038/nsmb.2007, PMID: 21336276

Hofer A, Crona M, Logan DT, Sjöberg BM. 2012. DNA building blocks: keeping control of manufacture. *Critical Reviews in Biochemistry and Molecular Biology* **47**: 50–63. DOI: https://doi.org/10.3109/10409238.2011.630372, PMID: 22050358

Holm L. 2022. Dali server: structural unification of protein families. *Nucleic Acids Research* **50**: W210–W215. DOI: https://doi.org/10.1093/nar/gkac387, PMID: 35610055

James El, Murphree TA, Vorauer C, Engen JR, Guttman M. 2022. Advances in hydrogen/deuterium exchange mass spectrometry and the pursuit of challenging biological systems. *Chemical Reviews* **122**: 7562–7623. DOI: https://doi.org/10.1021/acs.chemrev.1c00279, PMID: 34493042

Johansson R, Jonna VR, Kumar R, Nayeri N, Lundin D, Sjöberg BM, Hofer A, Logan DT. 2016. Structural mechanism of allosteric activity regulation in a ribonucleotide reductase with double ATP cones. *Structure* **24**: 906–917. DOI: https://doi.org/10.1016/j.str.2016.03.025, PMID: 27133024

Jonna VR, Crona M, Rofougaran R, Lundin D, Johansson S, Brännström K, Sjöberg BM, Hofer A. 2015. Diversity in overall activity regulation of ribonucleotide reductase. *Journal of Biological Chemistry* **290**: 17339–17348. DOI: https://doi.org/10.1074/jbc.M115.649624, PMID: 25971975

Jumper J, Evans R, Pritzel A, Green T, Figurnov M, Ronneberger O, Tunyasuvunakool K, Bates R, Žídek A, Potapenko A, Bridgland A, Meyer C, Kohl SAA, Ballard AJ, Cowie A, Romera-Paredes B, Nikolov S, Jain R, Adler J, Back T, *et al.* 2021. Highly accurate protein structure prediction with AlphaFold. *Nature* **596**: 583–589. DOI: https://doi.org/10.1038/s41586-021-03819-2, PMID: 34265844

Kang G, Taguchi AT, Stubbe J, Drennan CL. 2020. Structure of a trapped radical transfer pathway within a ribonucleotide reductase holocomplex. *Science* **368**: 424–427. DOI: https://doi.org/10.1126/science.aba6794, PMID: 32217749

Kaufman SL, Skogen JW, Dorman FD, Zarrin F, Lewis KC. 1996. Macromolecule analysis based on electrophoretic mobility in air:globular proteins. *Analytical Chemistry* **68**: 1895–1904. DOI: https://doi.org/10.1021/ac951128f, PMID: 21619100

Larsson KM, Andersson J, Sjöberg BM, Nordlund P, Logan DT. 2001. Structural basis for allosteric substrate specificity regulation in anaerobic ribonucleotide reductases. *Structure* **9**: 739–750. DOI: https://doi.org/10.1016/s0969-2126(01)00627-x, PMID: 11587648

Liebschner D, Afonine PV, Baker ML, Bunkóczi G, Chen VB, Croll TI, Hintze B, Hung L-W, Jain S, McCoy AJ, Moriarty NW, Oeffner RD,

Poon BK, Prisant MG, Read RJ, Richardson JS, Richardson DC, Sammito MD, Sobolev OV, Stockwell DH, *et al.* 2019. Macromolecular structure determination using X-rays, neutrons and electrons: recent developments in *Phenix. Acta Crystallographicaw Section D Structural Biology* **75**: 861–877. DOI: https://doi.org/10.1107/S2059798319011471, PMID: 31588918

Logan DT, Andersson J, Sjöberg BM, Nordlund P. 1999. A glycyl radical site in the crystal structure of A class III ribonucleotide reductase. *Science* **283**: 1499–1504. DOI: https://doi.org/10.1126/science.283.5407.1499, PMID: 10066165

Lundahl MN, Sarksian R, Yang H, Jodts RJ, Pagnier A, Smith DF, Mosquera MA, van der Donk WA, Hoffman BM, Broderick WE, Broderick JB. 2022. Mechanism of radical S-Adenosyl-l-methionine adenosylation: radical intermediates and the catalytic competence of the 5'-deoxyadenosyl radical. *Journal of the American Chemical Society* **144**: 5087–5098. DOI: https://doi.org/10.1021/jacs.1c13706, PMID: 35258967

Lundin D, Berggren G, Logan DT, Sjöberg BM. 2015. The origin and evolution of ribonucleotide reduction. *Life* **5**: 604–636. DOI: https://doi.org/10.3390/life5010604, PMID: 25734234

Martínez-Carranza M, Jonna VR, Lundin D, Sahlin M, Carlson LA, Jemal N, Högbom M, Sjöberg BM, Stenmark P, Hofer A. 2020. A ribonucleotide reductase from clostridium botulinum reveals distinct evolutionary pathways to regulation via the overall activity site. *Journal of Biological Chemistry* **295**: 15576–15587. DOI: https://doi.org/10.1074/jbc.RA120.014895, PMID: 32883811

Mathews CK. 2016. The most interesting enzyme in the world. *Structure* **24**: 843–844. DOI: https://doi.org/10.1016/j.str.2016.05.006, PMID: 27276424

Mathews CK. 2018. Still the most interesting enzyme in the world. *FASEB Journal* **32**: 4067–4069. DOI: https://doi.org/10.1096/fj.201800790R, PMID: 30059250

Mulliez E, Ollagnier S, Fontecave M, Eliasson R, Reichard P. 1995. Formate is the hydrogen donor for the anaerobic ribonucleotide reductase from *Escherichia coli. PNAS* **92**: 8759–8762. DOI: https://doi.org/10.1073/pnas.92.19.8759, PMID: 7568012

Nii T, Maeda Y, Motooka D, Naito M, Matsumoto Y, Ogawa T, Oguro-Igashira E, Kishikawa T, Yamashita M, Koizumi S, Kurakawa T, Okumura R, Kayama H, Murakami M, Sakaguchi T, Das B, Nakamura S, Okada Y, Kumanogoh A, Takeda K. 2023. Genomic repertoires linked with pathogenic potency of arthritogenic*Prevotella copri*isolated from the gut of patients with rheumatoid arthritis. *Annals of the*

Rheumatic Diseases **82**: 621–629. DOI: https://doi.org/10.1136/ard-2022-222881, PMID: 36627170

Ormö M, Sjöberg BM. 1990. An ultrafiltration assay for nucleotide binding to ribonucleotide reductase. *Analytical Biochemistry* **189**: 138–141. DOI: https://doi.org/10.1016/0003-2697(90)90059-i, PMID: 2278383

Perez-Riverol Y, Bai J, Bandla C, García-Seisdedos D, Hewapathirana S, Kamatchinathan S, Kundu DJ, Prakash A, Frericks-Zipper A, Eisenacher M, Walzer M, Wang S, Brazma A, Vizcaíno JA. 2022. The PRIDE database resources in 2022: A hub for mass spectrometry-based proteomics evidences. *Nucleic Acids Research* **50**: D543–D552. DOI: https://doi.org/10.1093/nar/gkab1038, PMID: 34723319

Pettersen EF, Goddard TD, Huang CC, Meng EC, Couch GS, Croll TI, Morris JH, Ferrin TE. 2021. UCSF ChimeraX: Structure visualization for researchers, educators, and developers. *Protein Science* **30**: 70–82. DOI: https://doi.org/10.1002/pro.3943, PMID: 32881101

Punjani A, Rubinstein JL, Fleet DJ, Brubaker MA. 2017. cryoSPARC: algorithms for rapid unsupervised cryo-EM structure determination. *Nature Methods* **14**: 290–296. DOI: https://doi.org/10.1038/nmeth.4169, PMID: 28165473

Ramírez-Aportela E, Mota J, Conesa P, Carazo JM, Sorzano COS. 2019. *DeepRes*: a new deep-learning- and aspect-based local resolution method for electron-microscopy maps. *IUCrJ* **6**: 1054–1063. DOI: https://doi.org/10.1107/S2052252519011692, PMID: 31709061

Rofougaran R, Crona M, Vodnala M, Sjöberg BM, Hofer A. 2008. Oligomerization status directs overall activity regulation of the *Escherichia coli* class la ribonucleotide reductase. *The Journal of Biological Chemistry* **283**: 35310–35318. DOI: https://doi.org/10.1074/jbc.M806738200, PMID: 18835811

Rosenthal PB, Henderson R. 2003. Optimal determination of particle orientation, absolute hand, and contrast loss in single-particle electron cryomicroscopy. *Journal of Molecular Biology* **333**: 721–745. DOI: https://doi.org/10.1016/j.jmb.2003.07.013

Rozman Grinberg I, Lundin D, Hasan M, Crona M, Jonna VR, Loderer C, Sahlin M, Markova N, Borovok I, Berggren G, Hofer A, Logan DT, Sjöberg B-M. 2018a. Novel ATP-cone-driven allosteric regulation of ribonucleotide reductase via the radical-generating subunit. *eLife* **7**: e31529. DOI: https://doi.org/10.7554/eLife.31529, PMID: 29388911

Rozman Grinberg I, Lundin D, Sahlin M, Crona M, Berggren G, Hofer A, Sjöberg BM. 2018b. A glutaredoxin domain fused to the radical-generating subunit of ribonucleotide reductase (RNR) functions as an efficient RNR reductant. *Journal of Biological Chemistry* **293**:

15889–15900. DOI: https://doi.org/10.1074/jbc.RA118.004991, PMID: 30166338

Rozman Grinberg I, Martínez-Carranza M, Bimai O, Nouaïria G, Shahid S, Lundin D, Logan DT, Sjöberg BM, Stenmark P. 2022. A nucleotide-sensing oligomerization mechanism that controls NrdR-dependent transcription of ribonucleotide reductases. *Nature Communications* **13**: 2700. DOI: https://doi.org/10.1038/s41467-022-30328-1, PMID: 35577776

Sanchez-Garcia R, Gomez-Blanco J, Cuervo A, Carazo JM, Sorzano COS, Vargas J. 2021. DeepEMhancer: a deep learning solution for cryo-EM volume post-processing. *Communications Biology* **4**: 874. DOI: https://doi.org/10.1038/s42003-021-02399-1, PMID: 34267316

Sievers F, Wilm A, Dineen D, Gibson TJ, Karplus K, Li W, Lopez R, McWilliam H, Remmert M, Söding J, Thompson JD, Higgins DG. 2011. Fast, scalable generation of high-quality protein multiple sequence alignments using Clustal Omega. *Molecular Systems Biology* **7**: 539. DOI: https://doi.org/10.1038/msb.2011.75, PMID: 21988835

Stourac J, Vavra O, Kokkonen P, Filipovic J, Pinto G, Brezovsky J, Damborsky J, Bednar D. 2019. Caver Web 1.0: identification of tunnels and channels in proteins and analysis of ligand transport. *Nucleic Acids Research* **47**: W414–W422. DOI: https://doi.org/10.1093/nar/gkz378, PMID: 31114897

Torrents E, Buist G, Liu A, Eliasson R, Kok J, Gibert I, Gräslund A, Reichard P. 2000. The Anaerobic (Class III) Ribonucleotide Reductase from Lactococcus lactis. *Journal of Biological Chemistry* **275**: 2463–2471. DOI: https://doi.org/10.1074/jbc.275.4.2463

Torrents E, Eliasson R, Wolpher H, Gräslund A, Reichard P. 2001. The anaerobic ribonucleotide reductase from lactococcus lactis. *Journal of Biological Chemistry* **276**: 33488–33494. DOI: https://doi.org/10.1074/jbc.M103743200

Torrents E, Westman M, Sahlin M, Sjöberg BM. 2006. Ribonucleotide reductase modularity: atypical duplication of the ATP-cone domain in *Pseudomonas aeruginosa. The Journal of Biological Chemistry* **281**: 25287–25296. DOI: https://doi.org/10.1074/jbc.M601794200, PMID: 16829681

Vagin A, Teplyakov A. 1997. *MOLREP*: an automated program for molecular replacement. *Journal of Applied Crystallography* **30**: 1022–1025. DOI: https://doi.org/10.1107/S0021889897006766

Wei Y, Funk MA, Rosado LA, Baek J, Drennan CL, Stubbe J. 2014. The class III ribonucleotide reductase from Neisseria bacilliformis can utilize thioredoxin as a reductant. *PNAS* **111**: E3756–E3765. DOI: https://doi.org/10.1073/pnas.1414396111, PMID: 25157154

Wheeler TJ, Clements J, Finn RD. 2014. Skylign: a tool for creating informative, interactive logos representing sequence alignments and profile hidden markov models. *BMC Bioinformatics* **15**: 7. DOI: https://doi.org/10.1186/1471-2105-15-7, PMID: 24410852

Wiley. 2001. Encyclopedia of life sciences. https://onlinelibrary.wiley.com/doi/book/10.1002/047001590X [Accessed May 30, 2001].DOI: https://doi.org/10.1002/047001590X

Zhang N, Yu X, Zhang X, D'Arcy S. 2021. HD-eXplosion: visualization of hydrogen-deuterium exchange data as chiclet and volcano plots with statistical filtering. *Bioinformatics* **37**: 1926–1927. DOI: https://doi.org/10.1093/bioinformatics/btaa892, PMID: 33079991

Chapter 8

Mathematical Supplement/Appendix

Here we shall elaborate various mathematical concepts that have been mentioned in the book. It can be read in tandem with the chapters or, even independently, since these are self sufficient.

1. Shift invariant linear filters

Refinement algorithms use such filters to smoothen (i.e. remove high spectral frequency noise from a digital image) the 3D structure by employing the same kernel (generally a Gaussian). Here we explain the meaning of shift invariance and linearity.

Shift invariance

This implies the invariance of a system under shifts. Because of it the number of independent degrees of freedom reduces by one. Consider a system of two particles defined by coordinates x_1 and x_2. Giving a shift of d implies that new coordinates of the same system (the system remains the same since it is shift invariant) are

$$x_1' = x_1 + d, \quad x_2' = x_2 + d \tag{1}$$

So that,

$$x_2' - x_1' = x_2 - x_1 \tag{2}$$

Thus the difference of coordinates is a constant. Hence there is only one independent coordinate and we may set any of x_1 or x_2 to zero.

A general definition of shift invariance is now given. Consider a system H defined by

$$H[f(x)] = g(x) \tag{3}$$

where $f(x), g(x)$ are some functions. The system is shift invariant if,

$$H[f(x + x_0)] = g(x + x_0) \tag{4}$$

where x_0 is a constant.

Example 1. Let

$$H[f(x)] = f(2x) = g(x)$$

Then

$$g(x + x_0) = f(2x + 2x_0)$$

Let

$$\omega(x) = f(x + x_0)$$

Note that $f(x+x_0)$ is only a function of x since x_0 is constant. Then

$$H[f(x + x_0)] = H[\omega(x)] = \omega(2x) = f(2x + x_0)$$

Hence

$$H[f(x + x_0)] \neq g(x + x_0)$$

Hence the system is not shift invariant.

Example 2. Take

$$H[f(x)] = (f(x))^2 = g(x)$$

Then

$$g(x + x_0) = (f(x + x_0))^2$$

Let

$$\omega(x) = f(x + x_0)$$

Then

$$H[f(x + x_0)] = H[\omega(x)] = (\omega(x))^2 = (f(x + x_0))^2$$

Thus

$$g(x + x_0) = H[f(x + x_0)]$$

Hence the system is shift invariant.

Linearity: Once again, before giving a general definition let us discuss some simple basic ideas. A linear function of a single variable is defined as,

$$y = f(x) = ax + b, \quad a \,\&\, b \text{ are constants}$$

which is an equation for a straight line.

A special property is that a small change of the independent variable (here x) is proportional to a small change of the function (here $f(x)$ or y) because,

$$y + \Delta y = a(x + \Delta x) + b$$

so that

$$\Delta y = a\Delta x$$

where $\Delta x, \Delta y$ are small changes. This proves our assertion. For a function of two variables.

$$y = f(x_1, x_2) = ax_1 + bx_2 + c$$

is again a linear function of x_1 and x_2.

Small changes are related by,

$$\begin{aligned} \Delta y &= a\Delta x_1, && \text{when } x_2 \text{ is fixed} \\ \Delta y &= b\Delta x_2, && \text{when } x_1 \text{ is fixed} \\ \Delta y &= a\Delta x_1 + b\Delta x_2, && \text{when both vary} \end{aligned}$$

Apart from revealing the proportionality of the small changes, these are found to be independent, i.e. change in x_2 is unaffected by a change in x_1 and vice-versa.

The two properties, i) independence and ii) proportionality help to provide a general definition of linearity in a system denoted by $H[f(x)]$. Linearity means,

$$H[\alpha f_1(x) + \beta f_2(x)] = \alpha H[f_1(x)] + \beta H[f_2(x)]$$

here α and β are some constants.

Example 3. Take

$$H[f(x)] = f(2x)$$

Then

$$\alpha H[f_1(x)] + \beta H[f_2(x)] = \alpha f_1(2x) + \beta f_2(2x)$$

Take

$$\omega(x) = \alpha f_1(x) + \beta f_2(x)$$

so that,

$$H[\alpha f_1(x) + \beta f_2(x)] = H[\omega(x)] = \omega(2x) = \alpha f_1(2x) + \beta f_2(2x)$$

Thus

$$\alpha H[f_1(x)] + \beta H[f_2(x)] = H[\alpha f_1(x) + \beta f_2(x)]$$

Hence the system is linear, although it is not shift invariant (see Example 1).

Example 4. Take

$$H[f(x)] = (f(x))^2$$

Then

$$\alpha H[f_1(x)] + \beta H[f_2(x)] = \alpha(f_1(x))^2 + \beta(f_2(x))^2$$

Taking

$$\omega(x) = \alpha f_1(x) + \beta f_2(x)$$

so that,

$$H[\alpha f_1(x) + \beta f_2(x)] = H[\omega(x)] = (\omega(x))^2 = (\alpha f_1(x) + \beta f_2(x))^2$$

Thus

$$H[\alpha f_1(x) + \beta f_2(x)] \neq \alpha H[f_1(x)] + \beta H[f_2(x)]$$

since,

$$(\alpha f_1(x) + \beta f_2(x))^2 \neq \alpha(f_1(x))^2 + \beta(f_2(x))^2$$

Hence the system is not linear, although it is shift invariant (see Example 2).

Comment: It is simple to verify that $H[f(x)] = f(x)$ is both linear and shift invariant.

Thus a regulariser like $H[f(x)] = f(x)$ qualifies as a shift invariant linear filter.

2. Physics and mathematics of Fourier analysis

Fourier transformations and their applications are an essential topic in various branches of structural biology. Before embarking on Fourier transformations, which is the subject of Apppendix 3, here we shall consider the distinctive role of Fourier analysis from a holistic point of view.

Cryo-EM provides a truly remarkable picture of the electron. The advent of direct electron detection devices has done away with scintillation counts. Electrons can now be detected directly instead of following the circuitous route of scintillation counters, where photons are ejected from an impact of electrons and their counting gives an idea of the number of electrons. The direct detection of electrons, as done now, is an illustration of the particle nature of electrons. However, in electron microscopy, the wave nature of the electron becomes more obvious. Basically, the principle of an electron microscope is similar to the usual optical microscope, except that electrons are used instead of photons (light).

Cryo-EM is therefore among the rare instances where an electron simultaneously displays particle and wave characteristics. Thus a proper explanation of this phenomenon must be based on quantum mechanics. It is here that Fourier analysis comes into play. But one may argue, since Fourier analysis was discovered about a century before quantum mechanics came into existence, how is this possible? As we shall see there is a deep connection between Fourier analysis and Heisenberg's uncertainty principle that helps in solving the wave-particle duality concept.

Consider a single electron moving with a particular velocity. The solution is a single wave covering the entire space with a well defined wavelength. According to Louis de Broglie, any particle is associated with a wave whose wavelength is given by Planck's constant divided by its momentum. The appearance of this constant implies the quantum nature of the relation.

Let us take the travelling wave representing the electron by its simplest possibility — a travelling "cosine" wave moving from minus infinity to plus infinity,

$$y = \cos x \tag{5}$$

This wave can represent classically any disturbance like water waves, sound waves or electrical impulses etc (see Fig. 1). However for the electron, it represents, quantum mechanically, a probability wave. Actually, probability is given by the modulus square of the amplitude. Hence the correct probability plot is given by the square of (5),

$$y^2 = \cos^2 x \tag{6}$$

This means that the electron can be anywhere (from minus infinity to plus infinity) on the x-axis (since the wave is travelling along that axis) with equal probability. However we know the momentum of the electron (using de Broglie's relation and the wavelength) exactly. This is a manifestation of Heisenberg's uncertainty relation. If we know the motion (or momentum) precisely we have no idea of its location (or position).

Let us next take the electron's motion to be represented by a sum of two travelling waves with different wavelengths. This is also

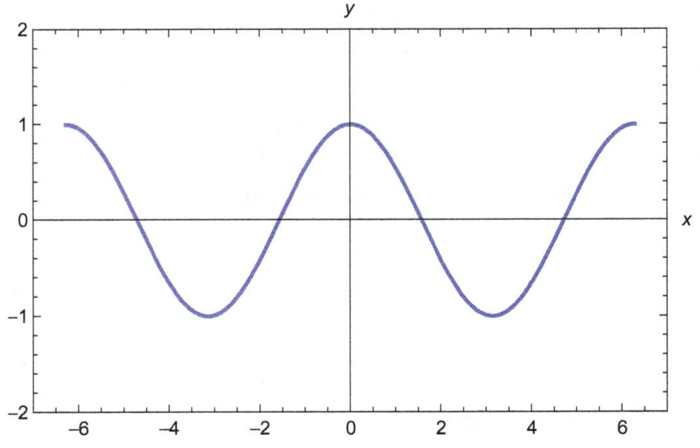

Figure 1: Plot of $y = \cos x$.

a travelling wave represented as,

$$y = y_1 + y_2, \quad y_1 = \cos x, \quad y_2 = \cos 2x \tag{7}$$

Graphically, we just add the two contributions. Since probability is given by squaring the amplitude we have to take the square of (7),

$$Y_2 = y^2 = (\cos x + \cos 2x)^2 \tag{8}$$

This is shown in Fig. 2. Now we see a localization effect. The electron has a greater probability of being found at the origin than at any other position. On the other hand the electron's momentum is no longer known completely accurately. There is a spread of two waves (or two wavelengths or two momenta).

If we consider the additions of more components (say 10), we find,

$$y = y_1 + y_2 + \cdots + y_{10}, \quad y_k = \cos kx, \quad k = 1, 2, \ldots 10 \tag{9}$$

The corresponding probability will be:

$$Y_{10} = \left(\sum_{k=1}^{10} y_k\right)^2 = \left(\sum_{k=1}^{10} \cos kx\right)^2 \tag{10}$$

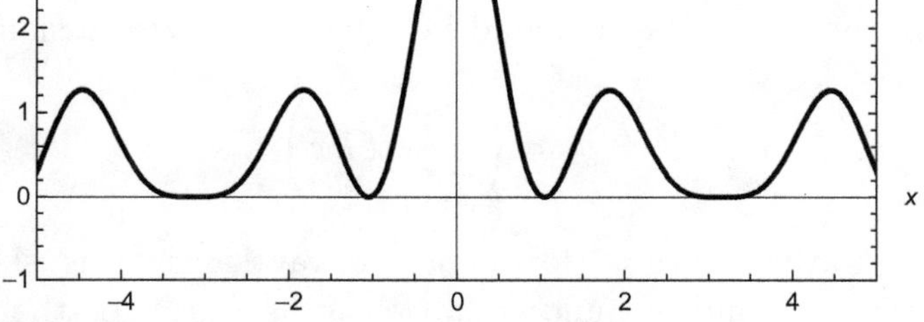

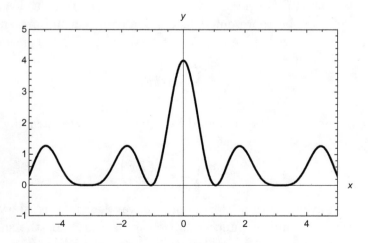

Figure 2: Plot of Y_2 (as defined in eqn (8)).

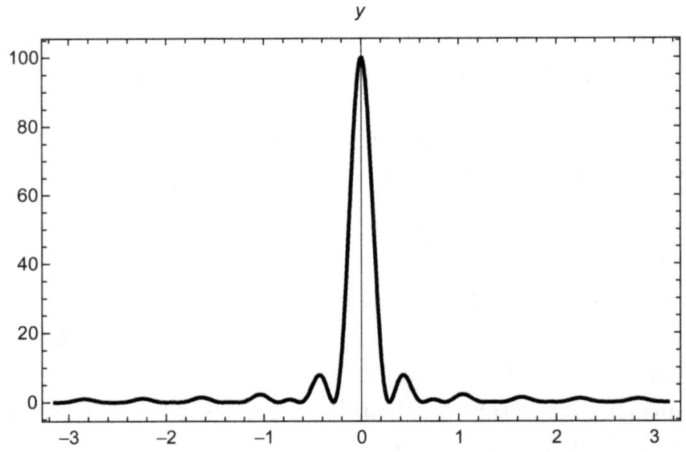

Figure 3: Plot of Y_{10} (as defined in eqn (10)).

Graphically, now we get the picture as shown in Fig. 3. Thus, as we increase the band with more and more waves, while the position gets fixed with increasing accuracy (the probability of the electron being at the origin is considerably greater than at any other position), its momentum gets more and more fuzzy. This is the essence of Heisenberg's uncertainty principle.

Finally, on to Fourier. His classic result is that a localised disturbance can be represented algebraically by an infinite sum of travelling waves, each with a different wavelength. Thus consider,

$$y = \sum_{k=1}^{\infty} \cos kx. \tag{11}$$

As usual, the probability is found by squaring the expression,

$$Y = y^2 = \left(\sum_{k=1}^{\infty} \cos kx \right)^2. \tag{12}$$

A representative picture (for hundred wavelengths) is shown in Fig. 4. With a band of hundred wavelengths (done for the sake of numerical illustration) the localisation of the particle at the origin in the Fig. 4 is almost complete. However, we know virtually next

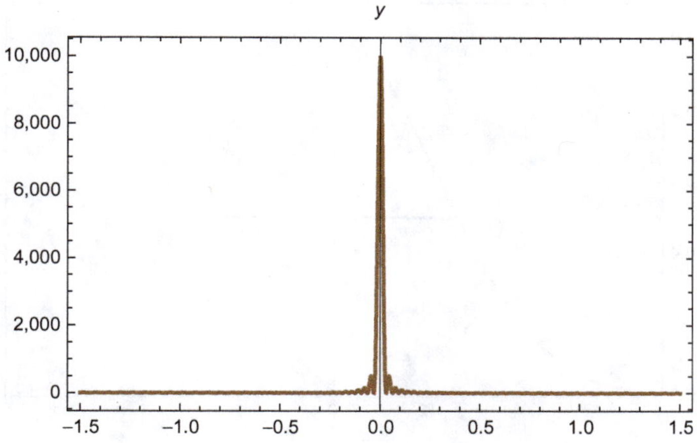

Figure 4: Plot of Y (for hundred wavelengths).

to nothing about its momentum since it is composed of hundred different wavelengths (or momenta).

The remarkable result of Fourier which shows the effect of localisation by taking a band of infinite waves is the key to understanding the dual concept of waves and particles. The wave nature of electrons in electron microscopy and the particle nature manifested in direct electron detectors are complementary aspects of the same reality seen through the analysis of Fourier.

3. Gaussian kernel and Fourier transform

Having learnt about Fourier analysis and its deep connection with quantum mechanics, here we consider Fourier transforms using Gaussian kernels which are frequently used in various barnches of structural biology, physics, engineering and mathematics.

Let us first explain what a Gaussian function is. It is defined as,

$$y = f(x) = e^{-x^2}, \tag{13}$$

which has the classic bell shaped curve shown in Fig. 5.

Since $x^2 \geq 0$ the maximum height of the curve is 1. Also, $f(x) = f(-x)$ implies that it is an even function of x. Hence the curve is symmetrical about the y-axis.

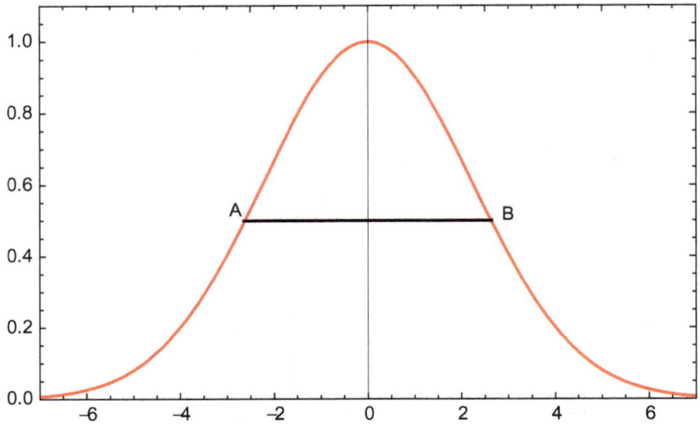

Figure 5: Plot of $y = e^{-x^2}$.

An important concept is the Gaussian width which is the distance between two symmetrical points (say A & B) on the curve. Of particular importance is the Gaussian width AB calculated at half length of the maximum height. In the figure this occurs at $x = \pm\sigma$. Then

$$\frac{1}{2} = e^{-\sigma^2} \Longrightarrow 1 = 2e^{-\sigma^2} \tag{14}$$

Taking logarithm on both sides,

$$0 = \log_e 2 - \sigma^2 \log_e e = \log_e 2 - \sigma^2 \Longrightarrow \sigma = \sqrt{\log_e 2} \tag{15}$$

Hence the Gaussian width at half maximum is,

$$2\sigma = 2\sqrt{\log_e 2} \tag{16}$$

Let us next calculate the integral,

$$I = \int_{-\infty}^{\infty} e^{-x^2} dx \tag{17}$$

which is the basic Gaussian integral. For future convenience, we take a general form,

$$I(a) = \int_{-\infty}^{\infty} e^{-ax^2} dx, \quad a > 0 \tag{18}$$

so that the integral does not diverge and is sensible. By squaring the integral,

$$I^2(a) = \int_{-\infty}^{\infty} e^{-ax^2} dx \int_{-\infty}^{\infty} e^{-ay^2} dy = \int_{-\infty}^{\infty}\int_{-\infty}^{\infty} dxdye^{-a(x^2+y^2)} \tag{19}$$

Introducing polar variables,

$$x = r\cos\theta, \quad y = r\sin\theta \tag{20}$$

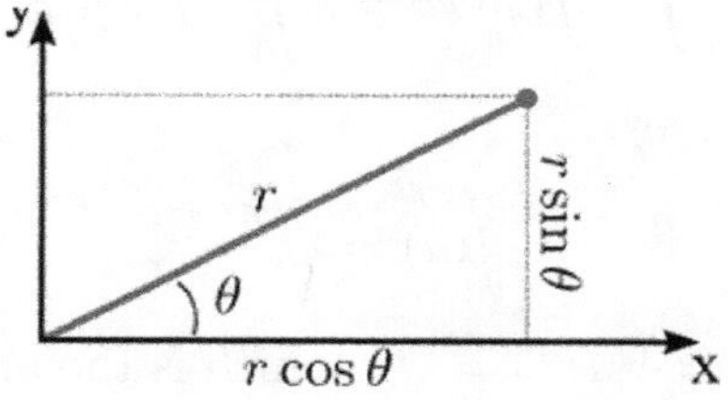

the integrand takes the form,

$$e^{-a(x^2+y^2)} = e^{-ar^2} \tag{21}$$

To obtain the integration measure we calculate first the differential elements,

$$dx = dr\cos\theta - r\sin\theta d\theta, \quad dy = dr\sin\theta + r\cos\theta d\theta \tag{22}$$

Since the measure defines the area enclosed by the differentials, we calculate their cross (or vector) product,

$$|dx \times dy| = |r\cos^2\theta dr \times d\theta| + |r\sin^2\theta dr \times d\theta| \tag{23}$$

where we have used,

$$dr \times dr = d\theta \times d\theta = 0, \quad dr \times d\theta = -d\theta \times dr \tag{24}$$

Hence,

$$|dx \times dy| = r|dr \times d\theta| \tag{25}$$

so that,

$$I^2(a) = \int_{-\infty}^{\infty} dx \int_{-\infty}^{\infty} dy e^{-a(x^2+y^2)} = 4\int_{0}^{\infty} dx \int_{0}^{\infty} dy e^{-a(x^2+y^2)}$$
$$= 4\int_{0}^{\infty} dr\, r \int_{0}^{\frac{\pi}{2}} d\theta e^{-ar^2} = 2\pi \int_{0}^{\infty} dr\, r e^{-ar^2} = \frac{\pi}{a} \tag{26}$$

In the final step the r-integral is solved by substituting $r^2 = t$. At an intermediate step we have used the fact that, for an even function,

$$\int_{-\infty}^{\infty} f(x)dx = 2\int_{0}^{\infty} f(x)dx \tag{27}$$

Hence the final answer is,

$$I(a) = \sqrt{\frac{\pi}{a}} \tag{28}$$

In particular, for $a = 1, I(1) = \sqrt{\pi}$ which is the area bounded by the curve shown in Fig. 5 and the x-axis,

$$\text{Area} = \int_{-\infty}^{\infty} e^{-x^2} dx = \sqrt{\pi} \tag{29}$$

From the result,

$$I(a) = \int_{-\infty}^{\infty} e^{-ax^2} dx = \sqrt{\frac{\pi}{a}} \tag{30}$$

we can derive other expressions by differentiation,

$$\frac{dI(a)}{da} = \int_{-\infty}^{\infty} (-x^2) e^{-ax^2} dx = -\frac{1}{2}\sqrt{\frac{\pi}{a^3}}$$
$$\Longrightarrow \int_{-\infty}^{\infty} x^2 e^{-ax^2} dx = \frac{1}{2}\sqrt{\frac{\pi}{a^3}} \tag{31}$$

By successive differentiations, the general expression for $\int_{-\infty}^{\infty} x^{2n}$ e^{-ax^2} may be evaluated,

$$\int_{-\infty}^{\infty} x^{2n} e^{-ax^2} dx = \left(\prod_{n=1}^{n} \left(\frac{2n-1}{2}\right)\right)\sqrt{\frac{\pi}{a^{2n+1}}} \tag{32}$$

$$\prod_{n=1}^{n} \left(\frac{2n-1}{2}\right) = \frac{1}{2} \times \frac{3}{2} \times \frac{5}{2} \cdots \tag{33}$$

A more general Gaussian has the form

$$f(x) = e^{-ax^2+bx} \tag{34}$$

which can be written, after completing squares,

$$f(x) = e^{-a\left(x-\frac{b}{2a}\right)^2} e^{\frac{b^2}{4a}} \tag{35}$$

Hence the Gaussian integral becomes,

$$I = \int_{-\infty}^{\infty} e^{-ax^2+bx} dx = \int_{-\infty}^{\infty} e^{-a\left(x-\frac{b}{2a}\right)^2} e^{\frac{b^2}{4a}} dx \tag{36}$$

Writing $x - \frac{b}{2a} = x'$

$$I = \int_{-\infty}^{\infty} e^{-ax'^2} dx' e^{\frac{b^2}{4a}} = \sqrt{\frac{\pi}{a}} e^{\frac{b^2}{4a}} \tag{37}$$

where eqn (30) has been used.

With this knowledge it is possible to discuss Fourier transforms with Gaussian functions (or kernels). Fourier transforms (F.T) are defined in pairs. Thus consider,

$$g(k) = \int_{-\infty}^{\infty} dx f(x) e^{-ikx}, \tag{38}$$

and

$$f(x) = \frac{1}{2\pi} \int_{-\infty}^{\infty} dk g(k) e^{ikx}, \tag{39}$$

Here $f(x)$ is the F.T of $g(k)$ and vice-versa. Alternatively, $f(x)$ and $g(k)$ are called "Fourier transform pairs".

Observe that while $f(x)$ is defined in the x-space (coordinate space), $g(k)$ is defined in the k-space (momentum space). There is nothing sacred about x-space and k-space being referred as coordinate space and momentum space. We may equally well denote $f(x)$ as an impulse response and $g(k)$ as a frequency response. These may also be referred as direct F.T and inverse F.T, depending on your choice. If there is a physical aspect, then $f(x)$ is defined in the real (physical) space while $g(k)$ is defined in the unphysical space. This is akin to the presence of direct (physical) and reciprocal (unphysical) lattice employed in crystallographic studies (see Appendix 6).

To see the internal consistency of the definitions (38) and (39), substitute (38) in (39) to get:

$$f(x) = \frac{1}{2\pi} \int_{-\infty}^{\infty} dk \int_{-\infty}^{\infty} dy f(y) e^{ik(x-y)} \tag{40}$$

Now

$$\frac{1}{2\pi} \int_{-\infty}^{\infty} dk e^{ik(x-y)} = \delta(x-y) \tag{41}$$

is the Dirac delta function which has the property that it vanishes everywhere except at $x = y$. It is similar to the localisation peak seen in Fig. 4 Appendix 2. Then,

$$f(x) = \int_{-\infty}^{\infty} dy f(y) \delta(x-y) = f(x) \tag{42}$$

thereby proving the consistency.

We now discuss F.T with Gaussian kernels. Taking the Gaussian its F.T is given by eqn (38),

$$g(k) = \int_{-\infty}^{\infty} dx e^{-ax^2 - ikx} dx \tag{43}$$

Putting $b = -ik$ and exploiting eqn (37), we obtain,

$$g(k) = \sqrt{\frac{\pi}{a}} e^{\frac{-k^2}{4a}} \tag{44}$$

which is another Gaussian. Thus the F.T of a Gaussian kernel yields another Gaussian. Note also that the parameter "a" in the Gaussian in eqn (36) appeared in the numerator of the exponential function. Its Fourier transformed version (44) has this parameter in the denominator. This reciprocity property among F.T. pairs is similar to the relation between distances among two planes in the direct and reciprocal lattice (see Appendix 6).

It is useful to reproduce the Gaussian (30) by taking the F.T of (44) using the definition (39),

$$f(x) = \frac{1}{2\pi} \int_{-\infty}^{\infty} dk \sqrt{\frac{\pi}{a}} e^{\frac{-k^2}{4a}} e^{ikx} \tag{45}$$

Using the eqns (36) and (37) the integral is worked out to yield the result,

$$f(x) = \frac{1}{2\pi}\sqrt{\frac{\pi}{a}}\sqrt{4a\pi}e^{-ax^2} = e^{-ax^2} \tag{46}$$

thereby reproducing the Gaussian function in eqn (30).

As a final remark we mention that, in several examples, performing a F.T. considerably simplifies the algebra. For instance if we consider the Dirac-delta function $\delta(x)$ and take its F.T. (38),

$$g(k) = \int_{-\infty}^{\infty} dx f(x)e^{-ikx} = \int_{-\infty}^{\infty} dx\delta(x)e^{-ikx} = 1 \tag{47}$$

since $\delta(x)$ only contributes when $x = 0$. Thus the F.T. of $\delta(x)$ is just unity.

4. Central slicing theorem

The central slicing theorem, also known as central section theorem or Fourier slicing theorem, is important in the reconstruction of objects from their images in one lower dimension. Thus, for a $2D$ reconstruction of an object, its projection in $1D$ is studied. There are two fundamental concepts that are necessary to understand this theorem. One is the analysis of Fourier and the other is the concept of projection. We have already dealt with the former. So we first explain projection before discussing the theorem.

The simplest example is the projection of a point on a line. In Fig. 6, the projection of P on AB is Q, the point where the perpendicular PQ cuts AB.

Its logical extension is the projection of a line on another line (Fig. 7). The projection of PQ on AB is MN which equals $PQ\cos\theta$, where θ is shown in the figure.

Angles affect projections. This is simply but effectively illustrated by projecting a beam of torch-light on a floor. If the torch is held perpendicular to the floor, the projected illumination is circular. If the torch is held inclined, the patch becomes elliptic. Holding the torch parallel to the floor, the torch beam fans out a parabola. By tilting further upwards we get a hyperbolic curve. The reconstruction of the $2D$ objects from their projection in $1D$ or that of $3D$ objects

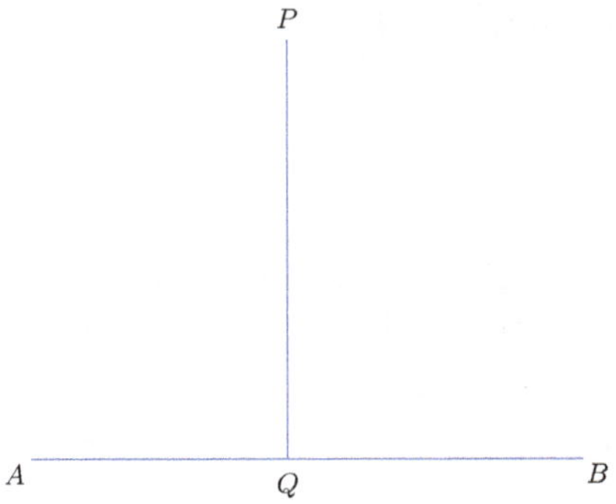

Figure 6: Pictorial representation of projection of a point P on a straight line AB.

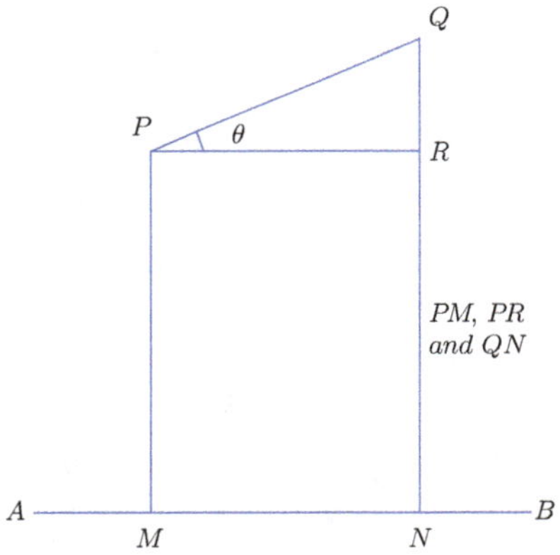

Figure 7: Pictorial representation of projection of a line PQ on another line AB.

from their $2D$ projections, therefore, depends crucially on the number of projections from various angles. Greater the number of projections, greater is the accuracy of the reconstruction. It is here that the central slicing theorem plays a pivotal role. We state and prove the theorem in $2D$ (i.e. two dimensions).

Statement of the theorem

The one dimensional Fourier transform of the projection of a two dimensional function is the same as the Fourier transform of the two dimensional function, after an appropriate slicing.

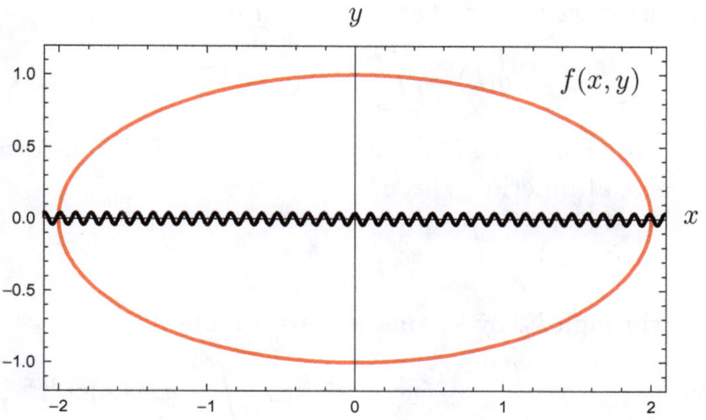

Figure 8: Projection of $f(x,y)$ on $x = p(x) = \int f(x,y)dy$.

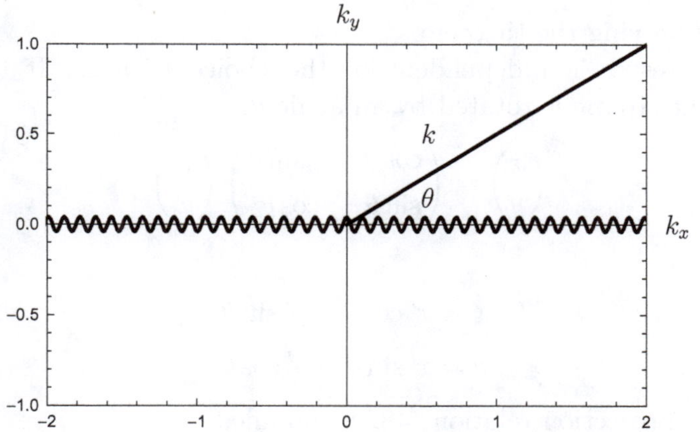

Figure 9: Corresponding Fourier transforms.

Proof. Consider a two dimensional function $f(x,y)$. Then its projection along the x-axis, $p(x)$ (see Fig. 8 and Fig. 9) is given by integrating out the y-coordinate,

$$p(x) = \int_{-\infty}^{\infty} f(x,y)dy \tag{48}$$

This is similar to the projection of PQ along the x-axis given by MN (7) which has no y-coordinate.

The Fourier transform of $p(x)$ is given by the formula,

$$q(k) = \int_{-\infty}^{\infty} dx p(x) e^{-ikx} \tag{49}$$

as follows from eqn (38). Now consider the Fourier transform of the two dimensional function $f(x, y)$,

$$g(k_x, k_y) = \int_{-\infty}^{\infty} dx \int_{-\infty}^{\infty} dy f(x, y) e^{-i(k_x x + k_y y)} \tag{50}$$

If we slice through k_y by setting it zero, we obtain,

$$g(k) = \int_{-\infty}^{\infty} dx \int_{-\infty}^{\infty} dy f(x, y) e^{-ikx} = \int_{-\infty}^{\infty} dx p(x) e^{-ikx} \tag{51}$$

where we have used eqn (48). Thus we find

$$g(k) = q(k) \tag{52}$$

thereby proving the theorem.

This result is independent of the choice of axes. If the x–y coordinate frame is rotated by an angle θ,

$$\begin{pmatrix} x \\ y \end{pmatrix} = \begin{pmatrix} \cos\theta & -\sin\theta \\ \sin\theta & \cos\theta \end{pmatrix} \begin{pmatrix} x' \\ y' \end{pmatrix} \tag{53}$$

so that,

$$\begin{aligned} x &= x' \cos\theta - y' \sin\theta, \\ y &= x' \sin\theta + y' \cos\theta \end{aligned} \tag{54}$$

then the projection relation (48) is modified to,

$$p(\theta, x') = \int_{-\infty}^{\infty} f(x' \cos\theta - y' \sin\theta, x' \sin\theta + y' \cos\theta) dy' \tag{55}$$

The $1D$ Fourier transform is now given by:

$$\begin{aligned} q(\theta, k) &= \int_{-\infty}^{\infty} p(\theta, x') e^{-ikx'} dx' \\ &= \int_{-\infty}^{\infty} \int_{-\infty}^{\infty} f(x' \cos\theta - y' \sin\theta, x' \sin\theta + y' \cos\theta) e^{-ikx'} dy' dx' \end{aligned} \tag{56}$$

The $2D$ Fourier transform of the function $f(x, y)$ is already defined in (50). In polar variables, corresponding to the angle θ (see Fig. 9),

$$k_x = k\cos\theta, \quad k_y = k\sin\theta \tag{57}$$

we obtain,

$$g(\theta, k) = \int_{-\infty}^{\infty} dx \int_{-\infty}^{\infty} dy f(x, y) e^{-i(k\cos\theta x + k\sin\theta y)} \tag{58}$$

Reverting back to the rotated coordinates (54),

$$g(\theta, k) = \int_{-\infty}^{\infty} dx' \int_{-\infty}^{\infty} dy' f(x'\cos\theta - y'\sin\theta, x'\sin\theta + y'\cos\theta) e^{-ikx'} \tag{59}$$

where we have used the fact,

$$dxdy = dx'dy' \tag{60}$$

since a rotation does not change the area element $(dxdy)$. Comparing (56) and (59),

$$q(\theta, k) = g(\theta, k) \tag{61}$$

thereby proving the theorem in the rotated frame.

In a similar manner it is possible to extend the theorem to higher dimensions.

Table 1 summarises the results of Fourier space analysis in $2D$ and $3D$.

Table 1: Fourier space analysis in $2D$ and $3D$

	$2D$	$3D$
Real space	$I(x, y)$	$\rho(x, y, z)$
What is this in EM?	an image	a structure (also called "map")
Function value is	Pixel intensity	The shielded Coulomb potential
Fourier space interpretation	We look at $\|F(s_x, s_y)\|^2$, the power spectrum, sometimes called EFT	Sometimes called molecular transform (knowledge of this yields the structure).

5. Eigenvectors, eigenvalues and principal component analysis

As mentioned in the text on cryo-EM, 3D variability analysis is based on determination of the eigenvectors of the 3D covariance of a set of particle images. This approach is also called the principal component analysis (PCA). For a proper understanding, concepts of eigenvectors and eigenvalues are essential. These are first introduced and then PCA is considered.

An eigenvalue equation is defined as

$$\hat{O}\vec{W} = \lambda\vec{W} \tag{62}$$

where $\hat{O}$ is an operator acting on an object $\vec{W}$ and λ is some number. Here $\vec{W}$ is called an eigenvector while λ is the eigenvalue.

The operator $\hat{O}$ could be a differential operator, a matrix or something abstract that is defined by certain properties. For example if $\hat{O}$ is the ordinary derivative operator $\frac{d}{dx}$ then the following equality

$$\frac{d}{dx}(e^{ax}) = a(e^{ax}) \tag{63}$$

is an eigenvalue equation. Here e^{ax} is an eigenfunction (of $\frac{d}{dx}$) with eigenvalue "a". For our purpose the relevant case occurs when the operator $\hat{O}$ is a matrix. Consider the following matrix equation

$$\begin{pmatrix} 0 & -1 \\ 2 & 3 \end{pmatrix} \begin{pmatrix} 1 \\ -2 \end{pmatrix} = 2\begin{pmatrix} 1 \\ -2 \end{pmatrix} \tag{64}$$

This is again an eigenvalue equation where the operator is represented by the 2×2 square matrix and the eigenvector is represented by the column matrix $\binom{1}{-2}$, and the eigenvalue is 2. The vector represented by $\binom{1}{-2}$ is obtained by joining the origin $(0, 0)$ with the point $(1, -2)$ in the $x - y$ plane (see Fig. 10). The relevant vector is $\overrightarrow{OP}$. The eigenvalue equation shows that the matrix operation changes just the magnitude of the vector (in this case, doubling it) but not the direction, leading to the vector $\overrightarrow{OQ}$ (Fig. 10). Equation (64) may, therefore be represented as,

$$\begin{pmatrix} 0 & -1 \\ 2 & 3 \end{pmatrix} \vec{OP} = 2\vec{OP} = \vec{OQ} \tag{65}$$

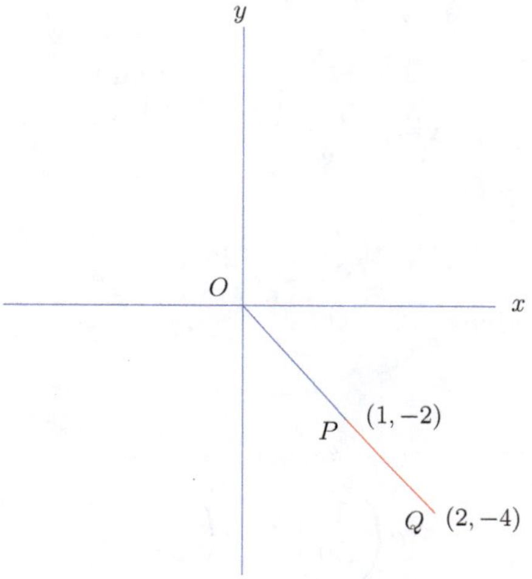

Figure 10: The vector representations of $\binom{1}{-2}$ and $\binom{2}{-4}$, both pointing in the same direction.

Thus directional flows of a vector are determined by such equations.

Now we discuss a method of finding the eigenvalues and the eigenvectors of a given matrix. Take the 2×2 matrix appearing in (64). In general a $n \times n$ matrix will have n eigenvalues and n eigenvectors. If the eigenvalues are non-degenerate (i.e. they are all unique) then all eigenvectors are independent, otherwise not. Thus the 2×2 matrix in (64) will have two eigenvalues that may or may not be degenerate. Denoting the eigenvalues by λ, these are given by solving the equation

$$\det \begin{pmatrix} 0 - \lambda & -1 \\ 2 & 3 - \lambda \end{pmatrix} = 0$$

$$\Longrightarrow \lambda^2 - 3\lambda + 2 = 0 \Longrightarrow \lambda = 2, 1 \tag{66}$$

Thus the eigenvalues of the 2×2 matrix in (64) are 2 and 1. Consider their eigenvectors to be given by $\binom{a}{b}$ so that,

$$\begin{pmatrix} 0 & -1 \\ 2 & 3 \end{pmatrix} \begin{pmatrix} a \\ b \end{pmatrix} = 2 \begin{pmatrix} a \\ b \end{pmatrix} \tag{67}$$

and

$$\begin{pmatrix} 0 & -1 \\ 2 & 3 \end{pmatrix} \begin{pmatrix} a \\ b \end{pmatrix} = 1 \begin{pmatrix} a \\ b \end{pmatrix} \tag{68}$$

The first of these yields

$$b = -2a \tag{69}$$

while the second gives

$$b = -a \tag{70}$$

Then the eigenvectors are given as

$$a\begin{pmatrix} 1 \\ -2 \end{pmatrix}, \quad a\begin{pmatrix} 1 \\ -1 \end{pmatrix} \tag{71}$$

One may fix a, by taking the length of the vector to be unity. This is allowed since changing the magnitude of the vector does not affect the eigenvalue equation. In the first case

$$a\begin{pmatrix} 1 & -2 \end{pmatrix} a\begin{pmatrix} 1 \\ -2 \end{pmatrix} = 1 \Longrightarrow 5a^2 = 1 \Longrightarrow a = \frac{1}{\sqrt{5}} \tag{72}$$

and likewise for the second case, that yields $a = \frac{1}{\sqrt{2}}$. Hence the final form of these (normalised) eigenvectors of the 2×2 matrix in (64) are,

$$\frac{1}{\sqrt{5}}\begin{pmatrix} 1 \\ -2 \end{pmatrix}, \quad \frac{1}{\sqrt{2}}\begin{pmatrix} 1 \\ -1 \end{pmatrix} \tag{73}$$

Before embarking on PCA, some statistical notions are introduced.

The average or mean value is defined as

$$n\bar{x} = \sum_{i=1}^{n} x_i \tag{74}$$

where the set x_i runs from $i = 1$ to $i = n$.

Variance, which is the square of the standard deviation, is defined as:

$$\text{var}(x) = \frac{\sum_{i=1}^{n}(x_i - \bar{x})(x_i - \bar{x})}{(n-1)} \tag{75}$$

while covariance among two variables x & y is defined as:

$$\text{cov}(x, y) = \frac{\sum_{i=1}^{n}(x_i - \bar{x})(y_i - \bar{y})}{(n-1)} \tag{76}$$

Note that var(x) is equal to the covariance among same variables

$$\text{var}(x) = \text{cov}(x, x) \tag{77}$$

Also note that

$$\text{cov}(x, y) = \text{cov}(y, x) \tag{78}$$

With this input it is possible to construct a 2×2 covariance matrix, which is symmetric

$$M = \begin{pmatrix} \text{var}(x) & \text{cov}(x, y) \\ \text{cov}(y, x) & \text{var}(y) \end{pmatrix} \tag{79}$$

While variance shows a departure from the mean (greater variance will have a greater spread), covariance shows a correlation between two variables. The sign is more important than the magnitude. A positive sign indicates a positive correlation: if x increases, y also increases or if x decreases y also does so. For a negative correlation, if x increases then y decreases and vice-versa. For zero correlation there is no relation between variation of x and y.

The eigenvalues and the eigenvectors of the covariance matrix carry useful information about data distribution. The eigenvectors are perpendicular and are interpreted as the principal components in PCA (principal component analysis). The eigenvalues indicate the magnitude of the spread in the direction of the principal components.

In PCA one retains the principal components with the large eigenvalues. This is an iterative process that successively eliminates eigenvectors with lower eigenvalues. This reduces the number of dimensions. Sacrificing accuracy to some extent, one attains an ease in computation.

6. Miller indices

Miller indices are used to define planes and their directions in a lattice or crystal. Hence these are frequently used in crystallographic studies.

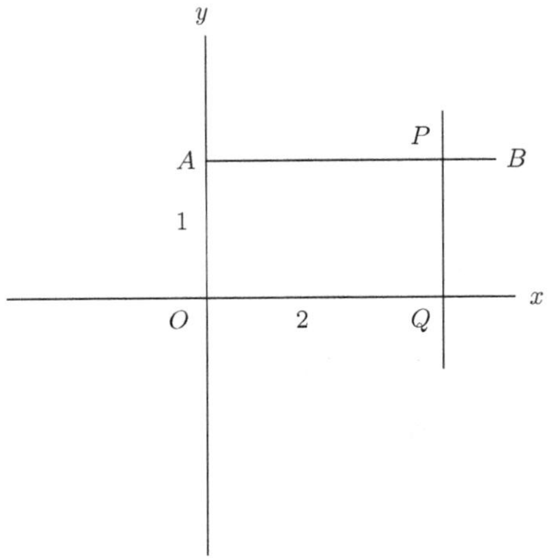

Figure 11: Line AP parallel to x-axis and the line PQ parallel to y-axis.

The principal advantage of these indices is their close connection to analytical geometry and removal of fractions by a suitable scaling or normalisation. They are best introduced by considering a planer surface or a two dimensional lattice. Take the following two lines: AB parallel to x axis and PQ parallel to y axis (Fig. 11). AB has an intercept of $OA = 1$ on the y axis. Its equation is given by,

$$y = 1 \tag{80}$$

and its intercept on the x axis is at ∞. The intercepts are denoted as $(\infty, 1)$. The reciprocal of it gives the Miller indices for AB,

$$\text{Miller indices} = (0, 1) \tag{81}$$

Likewise, the intercepts of PQ on x-axis and y-axis are UQ respectively, $OQ = 2$ and ∞ respectively. Their reciprocal is given by $(\frac{1}{2}, 0)$. Fractions are avoided by scaling (multiplying) the numbers by 2 to yield the Miller indices for PQ,

$$\text{Miller indices} = (1, 0) \tag{82}$$

At this point it is easy to realise how the Miller indices characterise the line and its direction. An entry "0" indicates that it is parallel to that axis while "1" indicates that it is perpendicular. Thus from the Miller indices of $AB = (0, 1)$, we find that AB is parallel to the x-axis and perpendicular to the y-axis which is consistent. The other thing to be noted is that, due the the normalisation trick, all lines parallel to AB will have Miller indices $(0, 1)$ and all lines parallel to PQ will have $(1, 0)$. Thus Miller indices correspond to a whole class or family of lines, instead of just a particular one.

Let us next consider a more general example, where the equation of a line is written in its intercept form or version (Fig. 12),

$$-\frac{x}{2} + \frac{y}{3} = 1 \tag{83}$$

This line has an intercept of -2 on the x-axis and 3 on the y-axis. The intercepts are denoted as $(-2, 3)$ whose reciprocal is $(-\frac{1}{2}, \frac{1}{3})$. Multiplying by 6 so that no fractions appear, we get the Miller indices of MN,

$$\text{Miller indices} = (\bar{3}, 2) \tag{84}$$

Note that the negative numbers are denoted by a bar over their head. Once again this represents a family of lines parallel to MN.

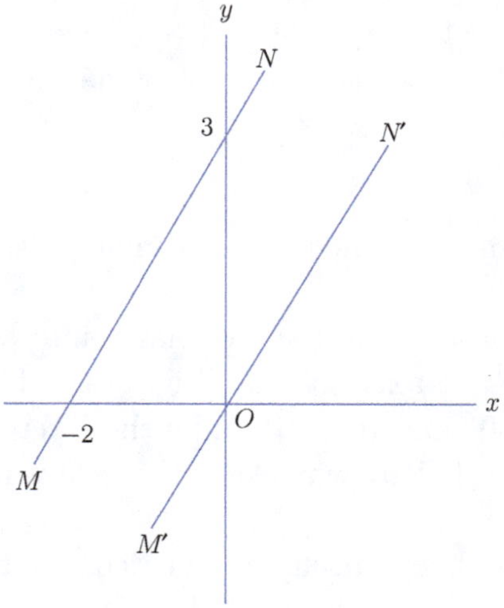

Figure 12: MN has intercepts -2 and 3 on x-axis, y-axis respectively. $M'N'$ is parallel to MN and passes through the origin O.

This is easily seen by modifying eqn (83) to:

$$-\frac{x}{2} + \frac{y}{3} = c \tag{85}$$

where c is a constant. Then the intercepts are $(-2c, 3c)$ and their reciprocal is $(-\frac{1}{2c}, \frac{1}{3c})$. The Miller indices are $(\bar{3}, 2)$. The eqn (85) represents a family of lines, one for each value of c, all being parallel to MN. It is useful to mention that finding the Miller indices of a line passing through the origin is tricky. Consider such a line $M'N'$ parallel to MN and passing through the origin O. Its intercepts are zero on both axes. Hence their reciprocal is not defined. However, since Miller indices hold for a whole family of parallel lines, these indices for $M'N'$ are same as those of MN, i.e. $(\bar{3}, 2)$.

These notions are easily extended to three dimensions. Now there will be three Miller indices. Take the equation of a plane in three dimensions in its intercept form,

$$\frac{x}{2} + \frac{y}{3} - \frac{z}{4} = 1 \tag{86}$$

The intercepts on x, y and z axes, respectively, are denoted as $(2, 3, -4)$. Their reciprocal is $(\frac{1}{2}, \frac{1}{3}, -\frac{1}{4})$. Hence multiplying by 12, the Miller indices are found to be $(6, 4, \bar{3})$. Once again this represents the Miller indices for a whole family of planes given by:

$$\frac{x}{2} + \frac{y}{3} - \frac{z}{4} = c \tag{87}$$

where c is some constant. The proof is similar to the two dimensional example.

As a last exercise we take the plane intercepting the x-axis at 1, y-axis at $\frac{1}{2}$, z-axis at ∞ (Fig. 13). The coordinates of P and Q are $(1, 0, 0)$ and $(0, \frac{1}{2}, 0)$ while the intercepts are $(1, \frac{1}{2}, \infty)$. Their reciprocal is $(1, 2, 0)$ which are the relevant Miller indices of the plane.

We have so far been discussing lattices in real or coordinate space. However, just as in the case of Fourier analysis, where Fourier transforms occur in pairs, here lattices also occur in pairs — real (or direct) lattice and the reciprocal lattice, defined in momentum space.

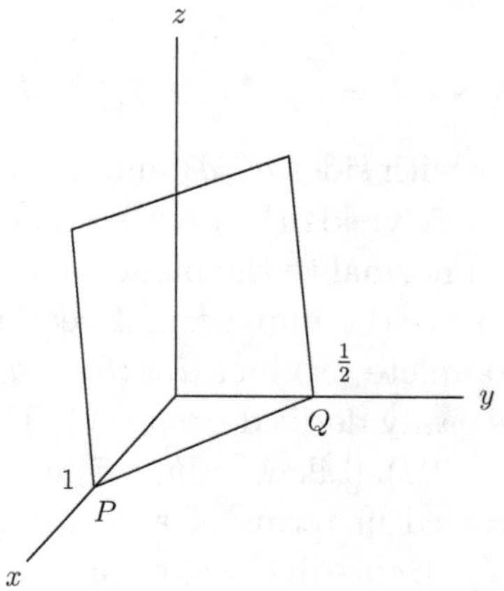

Figure 13: Plane with intercepts $(1, \frac{1}{2}, \infty)$.

Given a direct lattice there is a simple method of constructing the reciprocal lattice. Consider a radius vector in a direct lattice,

$$\vec{R} = \sum_{i=1}^{3} n_i \vec{a}_i = n_1 \vec{a}_1 + n_2 \vec{a}_2 + n_3 \vec{a}_3 \tag{88}$$

where a_1, a_2, a_3 are the basis (or primitive) vectors and n_1, n_2, n_3 are some integers. Then the radius vector in the corresponding reciprocal lattice is given by,

$$\vec{G} = \sum_{i=1}^{3} m_i \vec{b}_i = m_1 \vec{b}_1 + m_2 \vec{b}_2 + m_3 \vec{b}_3 \tag{89}$$

where b_1, b_2, b_3 are the basis vectors of the reciprocal lattice with m_1, m_2, m_3 some integers. Here $\vec{b}_i$ is defined in terms of $\vec{a}_i$ as,

$$\vec{b}_i \cdot \vec{a}_j = \delta_{ij}, \quad (i, j = 1, 2, 3) \tag{90}$$

where δ_{ij} is the Kronecker delta which is zero for $i \neq j$ and 1 for $i = j$. Thus,

$$\delta_{11} = \delta_{22} = \delta_{33} = 1, \quad \delta_{12} = \delta_{23} = \delta_{31} = 0 \tag{91}$$

Then $\vec{b}_i$ is written explicitly in terms of $\vec{a}_i$ as,

$$\vec{b}_1 = \frac{\vec{a}_2 \times \vec{a}_3}{V}, \quad \vec{b}_2 = \frac{\vec{a}_3 \times \vec{a}_1}{V}, \quad \vec{b}_3 = \frac{\vec{a}_1 \times \vec{a}_2}{V} \tag{92}$$

where,

$$V = \vec{a}_1 \cdot (\vec{a}_2 \times \vec{a}_3) = \vec{a}_2 \cdot (\vec{a}_3 \times \vec{a}_1) = \vec{a}_3 \cdot (\vec{a}_1 \times \vec{a}_2) \tag{93}$$

is the volume of a box with sides a_1, a_2 and a_3. The proof is easy. For the first equality, $\vec{a}_2 \times \vec{a}_3$ yields the area bounded by a_2 and a_3 with orientation (direction) normal to the plane containing $\vec{a}_2$ and $\vec{a}_3$. The scalar product of $\vec{a}_1$ with the unit normal vector gives the height of the box. Thus the complete product $\vec{a}_1 \cdot (\vec{a}_2 \times \vec{a}_3)$ yields V.

Relations (90) are easily derived from (92). The identity $\vec{b}_1 \cdot \vec{a}_1 = 1$ follows trivially from (92). Likewise $\vec{b}_1 \cdot \vec{a}_2 = 0$ follows by noting that $\vec{a}_2 \times \vec{a}_3$ is expressed in terms of a vector normal to the plane containing $\vec{a}_2$ and $\vec{a}_3$. Hence its scalar product with $\vec{a}_2$ vanishes. Similar derivations follow for other identities.

Further analysis will show that m_1, m_2, m_3 appearing in (89) are the Miller indices of planes orthogonal to $\vec{G}$ (i.e. the reciprocal lattice vector). These planes are defined on the direct lattice with intercepts $\frac{a_1}{m_1}, \frac{a_2}{m_2}, \frac{a_3}{m_3}$.

Finally, we mention that (92) is invertible, i.e. the vectors $\vec{a}_i$ can be expressed in terms of $\vec{b}_i$ as,

$$\vec{a}_1 = \frac{\vec{b}_2 \times \vec{b}_3}{\bar{V}}, \quad \vec{a}_2 = \frac{\vec{b}_3 \times \vec{b}_1}{\bar{V}}, \quad \vec{a}_3 = \frac{\vec{b}_1 \times \vec{b}_2}{\bar{V}} \tag{94}$$

where,

$$\bar{V} = \vec{b}_1 \cdot (\vec{b}_2 \times \vec{b}_3) = \vec{b}_2 \cdot (\vec{b}_3 \times \vec{b}_1) = \vec{b}_3 \cdot (\vec{b}_1 \times \vec{b}_2) \tag{95}$$

is now the volume of the box with sides b_1, b_2 and b_3. Relations (94) satisfy (90).

Ipsita: About Her

Sometimes a special, i.e. an out of the ordinary child is born who may not fit in the conventional graphic life of society, but they bear amazing abilities that may show miraculous results. Ipsita was one such girl. It is said that morning shows the day. Ipsita also started to show her deep admiration and adeptness to her studies when she was a mere child of only three years. Her peerless memory helped her to retain almost everything she read with passion. She had a natural skill for languages that made her a connoisseur in Japanese in a ten months schooling at Tsukuba, Japan as a class four student at the age of ten. Simultaneously she also became skillful in German language after a brief 'Kinderkurs' at Gothe Institute Kolkata.

An unconventional girl she was, grew up with innovative ideas from very childhood. To give a shape to an item from a reading material, was her natural gift that she began when she was a little girl of seven. As she grew up, gradually her attentiveness shifted more towards scientific experimentations. Picking up an item from some experimental book, she loved to remain engrossed and worked tirelessly to give it a final shape with tinkering materials. During that phase of work she literally used to forget her hunger and thirst and lifted her head only after making a perfect finish. These were the pastimes of her school days.

Her level of concentration was a yardstick to her talents and she loved to win and win only in whatever she did and though very compassionate at heart, got a great pleasure crushing down her opponents. For example, when she was a very little girl of one or two years her father sometimes, playfully lifting his hand, used to tell her that he would strike her with a 'Katari' (a sharp weapon) and spontaneously she replied that she would also give a 'cross-bolt' (probably she had heard that word from her father's mouth) in return in her babyish tone showing the same gesture as her father did. In her mind and simple little body she was a born conqueror and hated to be 'defeated'!

By the side of a great fighter, she was a dreamer also — that was the paradox of her life and characteristics. She loved to beautify everything she did (especially in her work), whether it be in experimentation or writing. On the contrary, she was very careless and totally indifferent to her own self and never admired herself in a mirror or used any cosmetics. Her world of beautification was her working sphere. Among all languages she had an expertise in English and she loved to make it lyrical. She often said, I hate the sacrilege to a language! She was a voracious reader and had a very wide range of reading books and knowledge as well. Actually her dreaming aptitude along with beautiful British English got a push from the Young Children's Encyclopedia set of Britannica she got as a gift from her parents at a tender age of six-seven. It was her best loved among others and it triggered and taught her to envisage her work through the fascinating beautiful corridor of a dream. It was the impetus that finally worked in her Ph. D. thesis of cryo-EM.

A mere Ph. D. student, with a very sketchy experimental background from India, she went to Lund to do experimental research in Biophysics following crystallography technique in which her guide excelled. But unfortunately that method failed repeatedly to bring success in her project. The years rolled by that brought her only despair, depressions, warnings from her Ph.D. committees and nothing else. In the meantime she had done a super-mini course on cryo-EM just for beginners, for six to seven days. One day, a flash of dream came to her while she was travelling to Stockholm carrying her dry-ice packed heavy bag of protein. She did this every month tirelessly without break, but with no result. In that journey, probably while working with her MacBook, she visualized her dream within cryo-EM. She could conjure up the beauty, the spark, the ambition, that could work as an impetus to fulfill her work — no matter in whatever magnitude that would be! She said that to her mother later on, with whom she loved to share every bit of her emotions and feelings. It worked as an inspiration that was in built in her instinct from the very beginning. In the meantime Covid closed everything and Ipsita took that opportunity to work hard and hard, and hard alone, without 'Any' help from outside (not also from her mentor, who was an enthusiast but was unaware of that technique) and finally made the glorious breakthrough.

The dreamer in her was further manifested through origami, the unique and old paper folding art of Japan, which she learnt at the age of ten while studying at Japan. She mastered it to an extent that her fascination continued till the end. Later, she directed her super skill to construct models of viruses (with whom she used to work in lab) through multiple manipulations of origami papers in a phenomenal way that surprised all her teachers and friends. She was very quick and dexterous to give them various shapes which she distributed to all she loved in her department smilingly.

She did huge work and brought atomic scale resolutions in her project that led to marvelous papers and completed her thesis work in only two and half years without proper guidance — creating a new record in the world history of thesis project.

She did stupendous work but got crushed with a very tender physique and delicate health as there was no support from outside and there she had to struggle hard changing four rented accommodations (as per rule of Lund University) and doing everything in a hard and harsh Nordic climate quite opposite to her own. That led her to get sick with a rare disease called SLE (Lupus) and after coming home she passed away on 7th December 2022, in the very month she was scheduled to defend her thesis.

Dr. Chirasree Banerjee

www.ingramcontent.com/pod-product-compliance
Lightning Source LLC
Chambersburg PA
CBHW050608010725
28797CB00004B/29

* 9 7 8 9 8 1 1 2 9 6 5 2 9 *